PRAISE FO

SPOOKY ACTION AT A DISTANCE

"A good science writer has to show us the fallible men and women who made the theory, and then show us why, after the human foibles are boiled off, the theory remains reliable. No well-tested scientific concept is more astonishing than the one that gives its name to a new book by the *Scientific American* contributing editor George Musser, *Spooky Action at Distance*. The ostensible subject is the mechanics of quantum entanglement; the actual subject is the entanglement of its observers. Musser presents the hard-to-grasp physics of 'non-locality,' and his question isn't so much how this weird thing can be true as why, given that this weird thing had been known about for so long, so many scientists were so reluctant to confront it."
—Adam Gopnik, *The New Yorker*

"In this polished study of the concept that Albert Einstein dubbed 'spooky action at a distance,' science writer George Musser tours the entangled research, history and philosophical speculation surrounding it . . . , proving that this is one of the most engrossing disputes in science."   —*Nature*

"Musser explores the history of humans grappling with nonlocality and what these strange effects are teaching quantum mechanics researchers, astronomers, cosmologists and more about how the universe works—and while doing so, showing the messy, nonlinear and fascinating way researchers push forward to understand the physical world."
—Sarah Lewin, Space.com

"With clever metaphors and dry humor, the acclaimed science communicator George Musser is the perfect tour guide on this wild ride through wormholes and emergent dimensions to the cutting edge of physics. This quest to understand the ultimate nature of space may forever transform how you think about the very fabric of reality."
—Max Tegmark, physicist and author of
Our Mathematical Universe

"Modern physics is in the process of dismantling the very space all around us, and the universe will never be the same. In this engaging book, George Musser leads us through the thickets of science and philosophy and takes us to the brink of a very different view of the world." —Sean Carroll, theoretical physicist at the California Institute of Technology and author of
The Particle at the End of the Universe

"Locality has been a fruitful and reliable principle, guiding us to the triumphs of twentieth-century physics. Yet the consequences of local laws in quantum theory can seem 'spooky' and nonlocal—and some theorists are questioning locality itself. *Spooky Action at a Distance* is a lively introduction to these fascinating paradoxes and speculations."
—Frank Wilczek, winner of the Nobel Prize in Physics and author of *The Lightness of Being* and *A Beautiful Question*

ADRIANNE MATHIOWETZ

GEORGE MUSSER
SPOOKY ACTION AT A DISTANCE

George Musser is an award-winning journalist, a contributing editor for *Scientific American*, and the author of *The Complete Idiot's Guide to String Theory*. He is the recipient of a Jonathan Eberhart Planetary Sciences Journalism Award from the American Astronomical Society and an American Institute of Physics Science Communication Award for Science Writing. He was a Knight Science Journalism Fellow at MIT and has appeared on *Today*, CNN, NPR, the BBC, Al Jazeera, and other outlets. He lives in Glen Ridge, New Jersey, with his wife and daughter. Follow him on Twitter at @gmusser and visit his website at www.georgemusser.com.

ALSO BY GEORGE MUSSER

The Complete Idiot's Guide to String Theory

SPOOKY ACTION
AT A DISTANCE

SPOOKY ACTION
AT A DISTANCE

SPOOKY ACTION AT A DISTANCE

The Phenomenon That Reimagines Space and Time—and What It Means for Black Holes, the Big Bang, and Theories of Everything

GEORGE MUSSER

Scientific American / Farrar, Straus and Giroux

New York

Scientific American / Farrar, Straus and Giroux
18 West 18th Street, New York 10011

An excerpt from *Spooky Action at a Distance* originally appeared,
in slightly different form, in *Scientific American*.

The Library of Congress has cataloged the hardcover edition as follows:
Musser, George.
    Spooky action at a distance : the phenomenon that reimagines
space and time—and what it means for black holes, the big bang, and
theories of everything / George Musser. — First edition.
        pages   cm
    Includes bibliographical references and index.
    ISBN 978-0-374-29851-7 (hardcover) — ISBN 978-0-374-71355-3 (e-book)
    1. Space and time—Philosophy.   2. Relativity (Physics)   I. Title.

QC173.59.S65 M88 2015
530.11—dc23

2015010155

Paperback ISBN: 978-0-374-53661-9

Designed by Jonathan D. Lippincott

Our books may be purchased in bulk for promotional, educational, or business use.
Please contact your local bookseller or the Macmillan Corporate and
Premium Sales Department at 1-800-221-7945, extension 5442, or by e-mail at
MacmillanSpecialMarkets@macmillan.com.

www.fsgbooks.com • books.scientificamerican.com
www.twitter.com/fsgbooks • www.facebook.com/fsgbooks

Scientific American is a registered trademark of Nature America, Inc.

5 7 9 10 8 6 4

For Talia and Eliana

Contents

Contents

SPOOKY ACTION
AT A DISTANCE

Introduction:
Einstein's Castle in the Air

When I first learned about nonlocality as a graduate student in the early 1990s, it wasn't from my quantum-mechanics professor: he didn't see fit to so much as mention it. Browsing a local bookshop, I picked up a newly published book, *The Conscious Universe*, which startled me with its claim that "no previous discovery has posed more challenges to our sense of everyday reality" than nonlocality. The phenomenon had the taste of forbidden fruit.

In everyday speech, "locality" is a slightly pretentious word for a neighborhood, town, or other place. But its original meaning, dating to the seventeenth century, is about the very concept of "place." It means that everything *has* a place. You can always point to an object and say, "Here it is." If you can't, that thing must not really exist. If your teacher asks where your homework is and you say it isn't anywhere, you have some explaining to do.

The world we experience possesses all the qualities of locality. We have a strong sense of place and of the relations among places. We feel the pain of separation from those we love and the impotence of being too far away from something we want to affect. And yet quantum mechanics and other branches of physics now suggest that, at a deeper level, there may be no such thing as place and no such thing as distance. Physics experiments can bind the fate of two particles together, so that they behave like a pair of magic coins: if you flip them, each will land on heads or tails—but always on the same side as its partner. They act in a coordinated way even though no force passes through the space between them. Those particles might zip off to opposite sides of the universe, and still they act in unison. These particles violate locality. They transcend space.

Evidently nature has struck a peculiar and delicate balance: under most circumstances it obeys locality, and it *must* obey locality if we are to exist, yet it drops hints of being nonlocal at its foundations. That tension is what I'll explore in this book. For those who study it, nonlocality is the mother of all physics riddles, implicated in a broad cross section of the mysteries that physicists confront these days: not just the weirdness of quantum particles, but also the fate of black holes, the origin of the cosmos, and the essential unity of nature.

For Albert Einstein, locality was one aspect of a broader philosophical puzzle: Why are we humans able to do science at all? Why is the world such that we can make sense of it? In a famous essay in 1936, Einstein wrote that the most incomprehensible thing about the universe is that it is comprehensible. At first glance, this statement itself seems incomprehensible. The universe is not a conspicuously rational place. It is wild and capricious, full of misdirection and arbitrariness, injustice and misfortune. Much of what happens defies reason (especially when romance or driving is involved). Yet against this backdrop of inexplicable happenings, the world's rules glow with reassuring regularity. The sun rises in the east. Things fall when you drop them. After the rain comes a rainbow. People go into physics out of a conviction that these are not just gratifying exceptions to the anarchy of life, but glimpses of an underlying order.

Einstein's point was that physicists really had no right to expect that. The world needn't have been orderly at all. It didn't have to abide by laws; under other circumstances, it might have been anarchic all the way down. When a friend wrote to ask Einstein what he'd meant by the comprehensibility remark, he wrote back, "*A priori* one should expect a chaotic world which cannot be grasped by the mind in any way."

Although Einstein said comprehensibility was a "miracle" we shall never understand, that didn't stop him from trying. He spent his entire professional life articulating exactly what it is about the universe that makes it make sense, and his thinking set the course of modern physics. He recognized, for example, that the inner workings of nature are highly symmetrical, looking the same if you view the world from a different angle. Symmetry brings order to the bewildering zoo of particles that physicists have found; entire species of particles are, in a sense, mirror images of one another. But among all the properties of the world that give us hope for understanding it, Einstein kept coming back to locality as *the* most important.

Locality is a subtle concept that can mean different things to different people. For Einstein, it had two aspects. The first he called "separability," which says that you can separate any two objects or parts of an object and consider each on its own, at least in principle. You can take your dining chairs and put each one in a different corner of the room. They will not cease to exist or lose any of their features—size, style, cushiness. The entire dining-room set derives its properties from the chairs that make it up; if each chair can seat one person, a set of four chairs can seat four people. The whole is the sum of its parts. The second aspect that Einstein identified is known as "local action," which says that objects interact only by banging into one another or recruiting some middleman to bridge the gap between them. Whenever a distance separates us from someone, we know we cannot have any effect on that person unless we cross the distance and touch, talk to, punch—somehow, make direct contact with—that person, or send someone or something to do it for us. Modern technology does not evade this principle; it merely recruits new intermediaries. A phone translates sound waves into electrical signals or radio waves that travel through wires or open space and then get translated back into sound on the other end. At every step of the way, something has to make direct contact with something else. If there is even a hairline crack in the wire, the message gets as far as a scream on an airless moon. Simply put, separability defines what objects are, and local action dictates what they do.

Einstein captured these principles in his theory of relativity. Specifically, relativity theory says that no material thing can move faster than light. Without such an ultimate speed limit, objects might move infinitely fast and distance would lose its meaning. All the forces of nature must wend their way laboriously through space, rather than leap across it in a single bound, as physicists used to suppose. Relativity theory thereby provides a measure of isolation among separated objects and ensures their mutual distinctness.

Depending on your frame of mind, relativity theory and the other laws of physics are either a satisfying deep order to the universe or a series of killjoy rules, like an authoritarian parent trying to take all the fun out of life. How great it would be to flap our arms and fly—but sorry, no can do. We could solve the world's problems by creating energy—oh, physics won't allow that, either; we can only convert one form of energy into another. And now comes locality, yet another draconian diktat, to spoil our dreams of faster-than-light starships and psychic powers. Locality dashes sports fans' eternal hope that, by crossing their fingers or bellowing some insightful comment from their armchairs, they might give their team an edge on the playing field. Unfortunately, if your team is losing and you're serious about wanting to help, you'll have to get up and go to the stadium.

Yet locality is for our own good. It grounds our sense of self, our confidence that our thoughts and feelings are our own. With all due respect to John Donne, every man *is* an island, entire of himself. We are insulated from one another by seas of space, and we should be grateful for it. Were it not for locality, the world would be magical—and not in a happy, Disneyesque way. As much as sports fans may wish they could sway the game from their living rooms, they should be careful what they wish for, because supporters of the opposing team would presumably have this power, too. Millions of couch potatoes across the land would strain to give their side some advantage, making the game itself meaningless—a contest of fans' wills rather than of talent on the field. Not just sports games, but the entire world would become hostile to us. In a world without locality, objects outside your body could reach inside without having to pass through your skin, and your body would lose its ability to control its internal condition. You would blend into your environment. And that is the very definition of death.

●

By focusing on locality as a crucial prerequisite to comprehending nature, Einstein crystallized two thousand years of philosophical and scientific thought. For ancient Greek thinkers such as Aristotle and

Democritus, locality made rational explanation possible. When objects can affect one another only by making direct contact, you can explain any event by giving a blow-by-blow account of "this hit that, which in turn knocked into *that*, which in turn bounced off some other thing." Every effect has a cause linked to it by a chain of events unbroken in space and time. There's no point at which you have to wave your hands and mumble, "Then a miracle occurs." It wasn't the miracle the Greek philosophers objected to—they weren't atheists—so much as the mumbling. Even gods, they felt, should exert their power by clear and explicable rules. Locality is essential not just to the types of explanations that philosophers and scientists seek, but to the methods they use. They can isolate objects from one another, grasp them one at a time, and build up a picture of the world step by step. They are not faced with the impossible task of taking it all in at once.

In 1948, toward the end of his life, Einstein summarized the importance of locality in a short essay: "The concepts of physics refer to a real external world . . . things that claim a 'real existence' independent of the perceiving subject . . . These things claim an existence independent of one another, insofar as these things 'lie in different parts of space.' Without such an assumption of the mutually independent existence . . . of spatially distant things, an assumption that originates in everyday thought, physical thought in the sense familiar to us would not be possible. Nor does one see how physical laws could be formulated and tested without such a clean separation."

Locality has such a pervasive importance because it is the essence of what space is. By "space" I don't just mean "outer space," the realm of astronauts and asteroids, but the space between us and all around us, the space that our bodies and everything else occupy, the space through which we swing a baseball bat or stretch a measuring tape. Whether you point your telescope at the planets or at the next-door neighbors, you are peering across space. For me, the beauty of a landscape comes from the giddy sense of spanning space, a sort of horizontal vertigo when you realize the little dots on the other side of a valley really are there and that you could touch them if only your arm were long enough.

As painters have long recognized, space is not mere absence, but a thing in its own right. What comes between objects on a canvas is as important to a composition as the objects themselves. For a physicist, space is the canvas of physical reality. Almost every attribute of our physical selves is spatial. We occupy a place. We have a shape. We move. Our bodies are intricate choreographies of cells and fluids dancing in space. Our thoughts are impulses zapping along pathways in space. Every interaction we have with the rest of the world passes through space. Living things are *things*, and what is a thing but a part of the universe that acquires an individual identity by virtue of occupying a certain volume of space?

Physics is rooted in the study of how things move through space, and space defines practically every quantity that physics deals in: distance, size, shape, position, speed, direction. Other qualities of the world may not appear spatial, but are; color, for example, corresponds to the size of a light wave. Only a very few properties of matter have no known spatial explanation, such as electric charge, and even these betray themselves by deflecting motion through space. When we look at an object, everything about it is ultimately spatial, arising from how its particles are arranged; the particles themselves are the barest flecks. Function follows form. Even nonspatial concepts become spatial in physicists' minds; time becomes an axis on a graph, and the laws of nature operate within abstract spaces of possibility. No less an authority than Immanuel Kant, whose ideas were a major influence on Einstein, thought it impossible to conceive of the world without space.

•

What a twist of fate that the greatest champion of locality was also its undoer. Though best known to the wider world for relativity theory, Einstein actually won his Nobel for cofounding quantum mechanics, the theory that describes how atoms and subatomic particles behave. Actually, physicists think quantum mechanics describes how *everything* behaves, although its distinctive effects are strongest on tiny scales. The theory grew out of Einstein's and his contemporaries' epiphany that atoms and particles can't just be littler versions of the things we see around us. If they were—if they acted according to the classical laws of physics developed by Isaac Newton and others—the world would self-destruct. Atoms would implode; particles would explode; lightbulbs would fry you with deadly radiation. The fact we're still alive means that matter must be governed by some new set of laws. Einstein welcomed the strangeness; in fact, despite the (unfair) reputation he later acquired as a rearguard defender of classical physics, he was consistently ahead of everyone else in appreciating the alien features of the quantum world.

Among those features was nonlocality. Quantum mechanics predicts that two particles can become blood brothers. For want of a mechanism to couple them, the particles should be completely autonomous, yet to touch one is to touch the other, as if distance meant nothing to them. The scientific method of divide and conquer fails for them. The particles have joint properties that escape you if you view them one at a time; you must measure the particles together. Our world is crisscrossed by a web of these seemingly mystical relationships. Atoms in your body retain a bond with every person you have loved—which sounds romantic until you realize that you're also linked to every weirdo who brushed against you while walking down the street.

Particles on opposite sides of the universe can't really be connected, can they? The idea struck Einstein as silly, a regression to prescientific notions of sorcery. Any theory that implied such "spooky actions at a distance," he reasoned, had to be missing something. He figured that the world was in fact local and merely gave the impression of being nonlocal, and he sought a deeper theory that would lay bare the hidden mechanism whereby two particles can act in unison. Try as he might, though, Einstein could never find such a theory, and he recognized that he might be the one who was missing something. There might be no concealed clockwork. The principle of locality—and with it, our conception of space—might not hold. A few months before he died, Einstein reflected on what the dissolution of space might mean for our understanding of the world: "Then *nothing* will remain of my whole castle in the air including the theory of gravitation, but also nothing of the rest of contemporary physics."

What was really spooky was how sanguine most of his contemporaries were. To them, nonlocality was a nonissue. The reasons for their dismissive attitude were complicated and are still debated by historians, but perhaps the most charitable explanation is pragmatism. The questions that vexed Einstein just didn't seem relevant to the practical applications of quantum theory. Only in the 1960s did a new generation of physicists and philosophers give Einstein's worries a real hearing. The experiments they did suggested that nonlocality was not a theoretical curiosity, but a fact of life. And even then, most of their colleagues gave it little thought—which is why I practically had to stumble on the topic as a grad student.

In the past twenty years, though, I've witnessed a remarkable evolution in attitudes. Nonlocality has surged into the currents of mainstream physics and swept far past the phenomenon that Einstein discovered. In my career as a science writer and editor, I have had the privilege of talking to scientists from a wide range of communities—people who study everything from subatomic particles to black holes to the grand structure of the cosmos. Over and over, I heard some variant of: "Well, it's weird, and I wouldn't have believed it if I hadn't seen it for myself, but it looks like the world has just got to be nonlocal." Researchers were like those matching particles on opposite sides of the universe, often not even knowing of one another, yet reaching the same conclusions.

If Einstein thought nonlocality smacked of sorcery, does the new research lend credence to paranormal claims? Some have thought so. In past decades, a number of scientists speculated that nonlocal links between particles could endow you with psychic powers. For instance, if particles in your brain were entangled with particles in your friend's, perhaps the two of you could communicate telepathically. At the other extreme, the supernatural intimations of nonlocality have been cause for many physicists to dismiss the whole area of research as hooey. In fact, there's no connection. None of the evidence for ESP has ever stood up, and the types of nonlocal phenomena under discussion are too subtle to meld minds or sway distant baseball games.

Some people are disappointed by that. They shouldn't be. The real magic of the world is that it *isn't* magical. For the reasons I discussed earlier, locality is a precondition for our existence. Any nonlocality must remain safely tucked away, emerging only under certain conditions, or else our universe would be inimical to life. What nonlocality gives us is much more impressive than any paranormal phenomenon: a window into the true nature of physical reality. If influences can leap across space as though it weren't really there, the natural conclusion is: *space isn't really there*. The Columbia University string theorist Brian Greene wrote in his 2003 book, *The Fabric of the Cosmos*, that nonlocal connections "show us, fundamentally, that space is not what we once thought it was." Well, what is it, then? Investigating nonlocality may clue us in. Many physicists now think that space and time are doomed—not fundamental elements of nature, but products of some primeval condition of spacelessness. Space is like a rug with ragged edges and worn spots. Just as we can look at those frayed areas to see how the rug is woven, we can study nonlocal phenomena to glimpse how space is assembled from spaceless components.

"I always thought, and still do, that the discovery and proof of the nonlocality is the single most astonishing discovery of twentieth-century physics," says Tim Maudlin, a professor at New York University and one of the world's leading philosophers of physics. In a paper in the late 1990s, he summed up the implications: "The world is not just a set of separately existing localized objects, externally related only by space and time. Something deeper, and more mysterious, knits together the fabric of the world. We have only just come to the moment in the development of physics that we can begin to contemplate what that might be."

At the same time, precisely because so much is at stake, other scientists tell me nonlocality can't be real—that one or another of the nonlocal phenomena will turn out to be a misinterpretation, and that it is a mistake to lump them together. Physicists have had enormous success with spatial reasoning and won't give it up lightly. One skeptic, Bill Unruh, who is a physics professor at the University of British Columbia, feels much as Einstein did: "If I have to know everything about the universe to know anything, if we take nonlocality seriously, if what happens here depends on what the stars are doing, it makes physics virtually impossible. What makes physics possible is that the world is partitionable. If we really do have to look to the stars to see our future, then I don't see how we can do physics anymore."

Apart from its inherent fascination, nonlocality is an ideal case study for scientific disputes. The disagreements between people such as Maudlin and Unruh are intellectually pure. No economic interests make you suspect ulterior motives. No lobbyists from Exxon-Mobil roam the halls. The adversaries have no overt personal animosity; many are friends. The mathematics is fairly simple; the experimental findings, undisputed. And still the debates drag on for generations. Today's scholars rehearse arguments that go back to Einstein and those he sparred with in the 1920s and '30s. Why is that? And what are the rest of us to do when the experts can't agree?

Consider the highest-profile scientific debate of recent times: climate change. Most climate scientists think human activity is warming the planet, some holdouts still disagree—and to someone reading a newspaper or surfing the web, the arguments can be baffling. Most people don't have time to become experts in general circulation models or measurements of longwave radiation. But one thing we will see is that a debate can be resolved in a practical sense regardless of whether the experts go on arguing. In the case of climate change, the public already knows what it needs to. There's a good chance of climate disaster and it's only prudent to manage the risk, just as you don't need a Ph.D. in combustion theory to know you should buy fire insurance for your home. Likewise, in the case of nonlocality, even the most die-hard skeptic now accepts that *something* very weird is going on, something that forces us to go beyond our deepest-held notions of space and time, something that we need to grasp if we are to know how the universe was born and how the natural world fits together in perfect unity.

The social stories are not just a sideline to the science. They are directly pertinent, because in a fluid area of research, where ideas jostle and nothing is entirely clear, the conventional ways that people outside science assume it operates—through the application of fact, logic, equations, experiments—aren't enough to bring closure. Scientists have to reach into gut feelings, metaphorical connections, and judgment calls about the adequacy of their basic principles. In deciding to explore nonlocality, I set off down what looked like a leisurely nature walk, but soon found myself entangled in an exotic rain forest, filled with glistening leaves, labyrinthine byways, and tempting handholds swarming with fire ants. Some physicists thrill to the rebelliousness of questioning one of the oldest and deepest concepts in science. Others shudder at the madness. If locality fails, does it mean our universe is ultimately incomprehensible, as Einstein feared, or can physicists find some other way for it to make sense?

The Many Varieties of Nonlocality

Enrique Galvez's lab at Colgate University is about the size of a two-car garage and, like most people's garages, jammed with stuff. Along the walls are workbenches loaded with toolboxes, electronic gear in various stages of disrepair, and, on the left side as you enter, the most frequently used piece of equipment: the coffee pot. In the middle of the room are a pair of optical benches: industrial-strength steel platforms, each the size of a dining-room table, covered with a pegboard-like grid of holes for attaching mirrors, prisms, lenses, and filters. "It's like playing with Erector sets all over again," says Galvez, a mellow Peruvian who looks remarkably like Al Franken.

If anyone has taken it on himself to show the world what quantum entanglement looks like, it's Galvez. Entanglement is the best known of several types of nonlocality that modern physicists have observed, and the one that spooked Einstein. The word "entanglement" has connotations similar to a romantic entanglement: a special and potentially troublesome relationship. Two particles that are entangled with each other are not literally intertwined, like balls of yarn; rather, they have a peculiar bond that transcends space. You can see this effect by creating, deflecting, and measuring beams of light—not ordinary flashlight beams, but beams of entangled photons. The earliest versions of the experiment, done in the 1970s at Berkeley and Harvard, involved mad-scientist contraptions of broiling-hot ovens, stacks of glass panes, and clattering teletypewriters. Galvez has taken advantage of Blu-ray lasers and optical fibers to miniaturize the setup, so that it now fits on a classroom desk.

Most experimental physicists I've met are tinkerers at heart, as fascinated by cool stuff as by the mysteries of the universe. An experimentalist at the Centre for Quantum Technologies in Singapore told me that, in his lab, incoming students have to pass a test. There's not a single physics question on it. Instead, they have to tell the story of how they took apart some household appliance and managed to get it back together, hopefully before their family found out. Apparently, clothes washers are a popular choice. Galvez, for his part, says his childhood passion was chemistry—of the blowing-up variety. Growing up in a middle-class neighborhood in Lima, he and some friends once tried to make gunpowder. All they got was a smoke bomb, which is perhaps just as well. "It was much more fun than something exploding," Galvez recalls. "It probably wasn't very healthy."

Galvez says he found his calling as a nonlocality crusader almost by accident. In common with the majority of physicists, he didn't give much thought to the phenomenon until the late 1990s, when a colleague stopped by his office with some dramatic news: the Austrian physicist Anton Zeilinger and his lab mates had used entanglement to teleport particles from one place to another. *Teleport?!* No fan of *Star Trek* could fail to be impressed. Although Zeilinger's team had beamed only single photons rather than an entire starship landing party, the coolness factor rivaled that of smoke bombs. And the procedure was straightforward. Suppose you want to teleport a photon from the left side of your lab to the right. First, you prime the teleporters by creating a pair of entangled photons and positioning one on each side of the lab. Then, you take the photon you want to beam and let it interact with the left particle. Because the entangled particles have a special bond between them, the interaction is immediately felt on the right, allowing the photon to be reconstituted there. (Some quibble whether the procedure should really be called teleportation; they consider it closer in spirit to identity theft. The experimentalists strip the left particle of its properties and thrust those properties onto the right particle. But a particle is nothing more than the sum of its properties, so these two characterizations amount to the same thing.)

Galvez and his colleague already had all the gear, and before long, they were beaming particles across their lab, too. "We were trying to figure out teleportation just for the fun of it," Galvez says. Another colleague suggested they design an entanglement experiment that even a physics-for-poets class could do. It doesn't do teleportation, but achieves the first and most important step in the process—namely, creating and distributing the entangled photons. As simple as the apparatus looks now, the team sweated over it for two years. Galvez began to run summer workshops for ALPhA, a physics-education group, to show teachers how to do the experiment, and he posted his instruction manuals online so that do-it-yourselfers can entangle particles in their basements. The former president of ALPhA, David Van Baak, exclaims: "We're past the stage where entanglement is a research-university-only affair. It's getting out to the masses."

On the day I visit Galvez's lab, one of his optical benches is given over to the entanglement experiment, the aim of which is not only to demonstrate entanglement, but also to explore what might be causing it. I recognize the setup as basically a high-tech Rube Goldberg coin flipper in which photons assume the role of coins. They are either "heads" or "tails" depending on whether they pass through a filter or not. The system is tuned so they have a 50-50 chance of getting through, like flipping a fair coin. The basic plan is to create a pair of these coins, flip both at the same time, see which sides they land on, create another pair, flip them, and so on. Repeat thousands of times and add up the statistics. It seems like a lot of effort for a predictable result, until you remember that we're talking about *quantum* coins. Clearly, thinking of particles as coins is a metaphor, but as long as you don't take it too literally, it's completely kosher. Physicists themselves understand phenomena in terms of metaphor.

To set the apparatus into motion, Galvez fires an ultraviolet laser through a series of optical elements that ensure proper alignment of the light. The beam strikes a small crystal of barium borate, a material discovered by Chinese scientists in the early 1980s, which splits the ultraviolet beam into two red beams. The splitting occurs particle

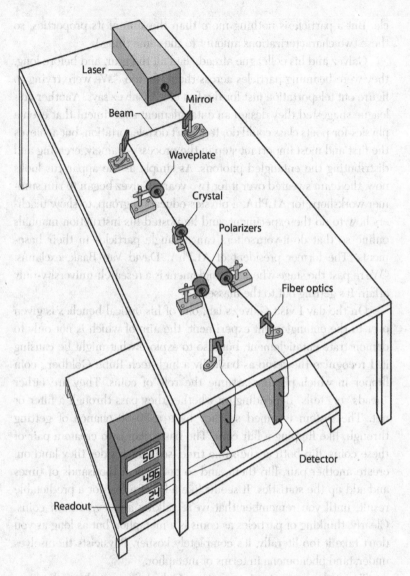

Laser

Mirror

Beam

Waveplate

Crystal

Polarizers

Fiber optics

Detector

507
496
24

Readout

1.1. Setup of quantum entanglement experiment. (Illustration by Jen Christensen) by particle in a process called downconversion: if you could zoom in and view the beam as a stream of photons, you would see some of the ultraviolet photons hit the crystal and divide their energy into identical twin red photons (see A Note on Entanglement). Voilà, coins. Located just upstream of the crystal is an optical element known as a waveplate, which Galvez uses to control the output of the crystal. Depending on how he sets the waveplate, the red photons are either entangled or not.

Once the red beams diverge, they cease to interact. Galvez aims each beam at a polarizing filter, much like the ones that photographers screw onto the front of their lenses to cut down on glare. The filter lets photons through or blocks them depending on their orientation—their polarization. Galvez can turn a dial on the side of the filter to control which photons make it. For this experiment, he sets both filters to the same angle, one that admits half the photons at random, thereby simulating coin flips.

Photons that make it through the filters are sent to detectors that convert them to electrical pulses. These detectors are the you-break-'em-you-bought-'em part of the system. Being sensitive enough to pick up individual photons, they run $4,000 apiece and are easily damaged by bright light. Even with the room lights off, the detectors pulse wildly, because the minutest sliver of light will set them off. Watching them gives me a new appreciation for how bright a supposedly dark room can be. We have to make sure our phones and laptops are fully off; a single glowing LED might spoil the experiment. "A while back we had to put black tape over anything that lit in the lab," Galvez says. "You would be surprised how many of those lights there are." He drapes a black velvet cloth over the devices and draws a thick curtain around the entire bench.

Finally, the detectors are wired into a meter with three digital readouts, located safely outside the curtain. Two show the number of photons that make it through the left and right polarizing filters. When Galvez switches on the laser, those numbers flash by like milliseconds on a stopwatch. The third readout shows the "coincidences"—when both photons in a pair make it through their respective filters. In terms of the coin metaphor, a coincidence means that both coins have landed on heads. These coincidences are Galvez's window into quantum nonlocality.

Having given me the tour, Galvez is ready to take some data. To verify that everything is working properly, he first simulates flipping ordinary coins by setting the waveplate to produce unentangled photons. The meter reads about twenty-five coincidences per second. For comparison, you'd get one hundred coincidences per second if every single photon in every single pair made it through the filters. So, the coincidence rate is about a quarter of its maximum possible value. This is just what you'd expect from the laws of chance. If you take two coins and flip them, each will come up heads about half the time, so both will be heads about a quarter of the time.

Now Galvez adjusts the waveplate to generate entangled photons. The coincidence rate jumps to about fifty per second. A change from twenty-five to fifty on a digital readout in a basement lab might not seem like much. But that's physics for you. It takes effort to peer beneath the surface of the world around us, and the clues are subtle, but they are no less dramatic for that. All those years of waiting and preparing for this moment have paid off, because when I see that fifty, I realize what I am seeing, and I shiver. The photons are behaving like a pair of magic coins. Galvez flips thousands of such pairs, and both always land on the same side: either both heads or both tails. That kind of thing doesn't happen by pure dumb luck.

Normal Coins		Quantum Entangled Coins	
Left Coin	Right Coin	Left Coin	Right Coin
heads	tails	heads	heads
tails	tails	tails	tails
heads	heads	heads	heads
tails	tails	tails	tails
tails	heads	tails	tails
tails	heads	tails	tails
heads	tails	heads	heads
heads	heads	heads	heads

1.2. Sample results of coins experiment. If you flip a pair of ordinary coins, they'll land on the same side half the time, on average. But if you flip a pair of suitably prepared quantum entangled "coins," they will *always* land on the same side.

If a friend of mine did this trick at a party—flip pairs of coins so that both came up heads twice as often as they rightfully should—I'd assume it was a practical joke. My friend might have gone to a magic shop and bought double-sided coins, which look the same on both sides, making the outcome of a flip preordained. Could an equivalent stunt explain the pattern I was seeing in Galvez's lab? To test for such trickery, Galvez uses a tactic proposed by the Irish particle physicist John Stewart Bell in the 1960s. He turns one of the filters by an angle of 90 degrees, which, like flipping a coin with your left hand rather than your right, doesn't alter the probability of a particle getting through; if the outcome really is predetermined, nothing should change. But this seemingly innocuous change *does* have an effect on the photons. The coincidence meter drops nearly to zero—meaning that if one photon gets through, the other *never* does. In other words, the magic coins have switched from always landing on the same side to always landing on opposite sides. A practical jokester would need some extra sleight of hand to pull off this trick. By making further refinements, Galvez rules out any conceivable chicanery.

I go over and look at the optical bench again. Those filters are separated by the width of my hand. Experiments by Zeilinger and others have stretched the distance to one hundred miles, and researchers at the Centre for Quantum Technologies are working on a space-based version that will go even farther. For a tiny particle, that might as well be the other side of the universe. The photons manage to coordinate their behavior across that gap. They are not in contact, and no known force links them, yet they act as one. When Galvez dials the polarizer filter on the left side of his lab bench and a photon passes through, the photon will be polarized in the same direction as the filter. Its entangled partner follows in lockstep: it acquires the same polarization and will respond accordingly to its own filter. So, what happens on the left affects the photon on the right, even when there's no time for any kind of influence to cross the gap. Indeed, such an influence would need to travel from left to right instantly—that is, infinitely fast, which is plainly faster than light, in apparent defiance of the theory of relativity. This is one of the many mysteries posed by nonlocality. Physicists have commented that it is as close to real magic as they've ever seen.

"Students love it," Galvez says. "The good students say, 'I want to figure this out.'"

Shut Up and Calculate

Is nonlocality just a carnival freak show—fun to ooh and aah over, but having no broader implications—or does it belong on the center stage of physics? For most of the twentieth century, physicists treated it as a freak show, and as a student I adopted this attitude, too. It wasn't until years later, when I delved into Tim Maudlin's book *Quantum Nonlocality and Relativity*, that I appreciated the depth of the mystery.

Sitting in his George Nakashima–furnished living room, Maudlin tells me he'll never forget the moment he learned about quantum nonlocality. One day in the fall of 1979, while a physics major at Yale University, he opened up the latest issue of *Scientific American* magazine. The cover story was about dung beetles, but he flipped past it and landed on an article on the early entanglement experiments. For particles to act as if by magic stunned Maudlin. "I remember the day when I read that article," he says. "My roommates remember that day. I walked around and around my room. The world wasn't what I thought it was. It bugged the hell out of me."

It also bugged him that his physics professors, like mine, had never once mentioned this phenomenon. When he probed them about it, they blew him off. Once, Maudlin recalls, he raised his hand in class and asked whether quantum theory might not give way to a deeper theory in which the seeming contradictions would make perfect sense. The professor dismissed the idea and went back to scribbling Greek letters on the blackboard. "He didn't offer any explanation at all of why not," Maudlin says. "So he shut down the question without answering it."

•

To appreciate the mind block that Maudlin and I ran into, you have to go back to the famous debates between Einstein and another of the founders of quantum mechanics, the Danish physicist Niels Bohr, in the 1920s and '30s. Einstein worried that nonlocality would contradict his theory of relativity and argued that it had to be a kind of illusion, reflecting our ignorance of some essential aspect of nature. Bohr argued . . . well, nobody is quite sure what Bohr argued. His reasoning gave "tangled" a whole new meaning, and his missives have been interpreted as either championing or contesting nonlocality. To the extent that anyone does understand what he said, he was asserting that it didn't matter what weirdness lay behind the scenes, as long as the theory could predict what experiments saw.

As anyone who has watched an American presidential debate knows, judgments about "win" or "lose" often have little to do with what the debaters actually say. Most physicists just wanted the Bohr-Einstein debate to be over, so they could get on with applying quantum mechanics to practical problems. Because Bohr promised closure, they rallied around him and wrote off Einstein as a has-been. One later wrote that Einstein's "fame would be undiminished, if not enhanced, had he gone fishing instead."

Over the subsequent decades, physicists used quantum theory to do all sorts of useful calculations. They figured out transistors, lasers, and other mainstays of the modern world. So the collective decision to set aside questions about the theory's deeper meaning seemed justified. Whenever those conceptual questions did come up, physicists deemed them "philosophical," which wasn't intended as a compliment, but as a way to deny that the questions were even worth asking. The English physicist Paul Dirac wrote, "It is only the philosopher, wanting to have a satisfying description of nature, who is bothered."

Because Maudlin was, in fact, bothered, he decided to go to graduate school in philosophy rather than physics. "I want to get to the bottom of everything," Maudlin says. "That's what you do as a philosopher." Philosophy is distinguished not just by its interests, but by its methods: philosophers are trained in logic as opposed to mathematics or experiments. Maudlin gained a reputation among philosophers as Dr. Takedown, able to spot the flaw in almost any argument. Throughout his graduate studies and early years as a professor, Maudlin says, nonlocality sizzled in the back of his mind. But no one else he knew seemed interested, and in some ways philosophers were as much in thrall to the principle of locality as physicists were. Other things kept getting in the way of Maudlin's thinking more about it—until the fall of 1990, when John Stewart Bell died.

Bell had done more than anyone else to reopen the case of Einstein versus Bohr. He began to doubt Bohr's victory while a university student in the early 1950s, but realized that airing his misgivings wouldn't do his career any favors. By the mid-'60s, having made a name for himself by studying particles and designing particle accelerators, including the precursors of the Large Hadron Collider, Bell felt secure enough to revisit his youthful concerns. He showed that nonlocality was no longer just a matter for debate; you could play with it in the lab. Like Einstein, Bell struggled to convince his colleagues. His first paper on the subject was not cited by a single other paper for four years and not mentioned in textbooks until 1985. Even when Bell's work did get attention, it was apt to be misinterpreted. One of his obituaries was titled "The Man Who Proved Einstein Was Wrong," which totally missed the man's point that nonlocality transcended the old debate. Einstein may have been wrong to think that nonlocality would prove to be merely apparent, but Bohr was wrong to have ignored it altogether.

Like Einstein, Bell fretted that nonlocality defied the theory of relativity. Physicists can't give up quantum theory; it passes all experimental tests. For relativity to be wrong is equally unthinkable. In a lecture in 1984, Bell concluded, "We have an apparent incompatibility, at the deepest level, between the two fundamental pillars of contemporary theory." Even those who were otherwise sympathetic to him saw no such incompatibility. In creating relativity theory, Einstein thought about how we gather information. Signals such as light or sound must pass from objects in the world to our senses. If those signals travel instantaneously, they can conflict. Paradoxes ensue. Things both happen and don't happen. The machinery of the universe seizes up. Yet quantum magic coins don't pose any such risk. They are inherently incapable of signaling. They land on either heads or tails—you can't dictate which. You have no way to control them in order to transmit a message or, indeed, do anything at all. So you could never use them to bring about a paradoxical situation. Danger averted.

In other words, if entanglement is magic, it is not like a magic wand that you can wave to make things happen. Rather, the magic happens spontaneously, and you notice it only if you're looking carefully. This is a very attenuated form of magic that won't win you any wizarding cups. Most everybody reassured themselves that quantum mechanics and the theory of relativity live in "peaceful coexistence."

In Bell's memory, several Rutgers University philosophers organized a symposium on quantum physics and asked Maudlin to speak. Picking up where he'd left off as a student, he proceeded to take down the lore that had grown up around Einstein's and Bell's findings. The conventional vision of theoretical harmony struck Maudlin as a trifle *too* harmonious. "Just pointing out that you can't send signals never seemed enough for me to show that there was no fundamental conflict with relativity," he says. Even if a pair of entangled particles can't convey a signal, quantum theory still says that what happens to one instantly affects the other. The theory therefore requires that the universe have a kind of master clock, ensuring that what is 7:30 p.m. to one particle is 7:30 p.m. to another. And relativity theory denies any such thing. The reason they call it *relativity* theory is that the passage of time is relative. Two events that are simultaneous for one person can be sequential for another.

Maudlin's book grew out of his talk, and its publication coincided with a surge of interest in entanglement. Experimentalists, realizing that the phenomenon wasn't as useless as they'd thought, were beginning to exploit it for cryptography and computers. For instance, Artur Ekert, a physicist at the University of Oxford and currently the director of the Centre for Quantum Technologies, proved in 1991 that entangled particles can create a communications channel so secure that not even the sneakiest government surveillance program could listen in. Once physicists were clued in to the importance of entanglement, they began to see it almost everywhere they looked. It occurs even in living organisms. In photosynthesis, entanglement accounts for the unexpectedly high efficiency with which molecules transfer light energy into chemical energy, thereby helping to enable life on our planet.

By the turn of the millennium, Einstein's paper that got it all started had become one of the most widely cited articles in the history of physics. Meanwhile, the old walls between physicists and philosophers were tumbling down. Zeilinger, the pioneering experimentalist, often disagrees with Maudlin, but exchanges ideas with him in a way that would have been unthinkable twenty years ago. "This connection between philosophy and physics is crucial to making real progress," Zeilinger tells me.

Quantum nonlocality is clearly not just a dinner act in Vegas, but an essential aspect of the world, and physicists and philosophers still don't know what is behind the magic. Could the clues they seek lie in other domains of science? What can they learn from the other types of nonlocality that are out there in the world?

The Skywatcher and the Ice Climber

For most of the twentieth century, the peculiar synchronicity of entangled particles was the only type of nonlocality that rated any mention. But physicists gradually realized that other phenomena are suspiciously spooky, too. Those who study black holes think that matter in these cosmic vacuum cleaners may jump from one place to another without crossing the intervening distance, a type of nonlocality that is arguably even more baffling than the situation Einstein worried about.

Black holes have long been physicists' top choice for weirdest things in the universe. Ramesh Narayan has seen them in action. Like Galvez, Narayan says he came to his scientific passion late and almost by accident. As a boy, he had zero interest in astronomy. He's one of the few astrophysicists I've met who doesn't recall being obsessed with black holes as a child. Crystals were his thing. But in his first job, at the prestigious Raman Research Institute in Bangalore, in southern India, he found himself mingling with people exploring the mysteries of the universe, and soon he was hooked. He became an expert on cosmic flows of gas. The governing principle of these flows is simple: what goes down must come up. Whenever gas crashes into the surface of a star, it acts to warm up the star; the star, in turn, emits the energy back into space, typically as infrared radiation or visible light. "All the energy falling in has to come out," explains Narayan, who is now a Harvard professor. But in the early 1990s astronomers noticed a peculiar exception to this rule at the center of our galaxy.

The center of the galaxy is easy to see for yourself. The next time you go outside to gaze at the night sky, find the constellation of Sagittarius. From my hometown, it is easiest to see in the summer and early fall, hanging in the sky near the south horizon. It is supposed to look like an archer, but most astronomers think of it as a giant teapot. The spout points to the very center of the Milky Way. To our eyes, the center is just a fuzzy patch of sky, but in the 1940s telescopes began to reveal a cauldron of swirling gas there. At the very center, gas converges on a tiny region known as Sagittarius A°. This region is puzzlingly dim; less than 1 percent of the energy carried by the inflowing gas comes back out. "In front of our eyes, you see the energy going into the center and vanishing—poof," Narayan says.

That is the definition of a black hole. Its gravity is so powerful that what goes down never comes up. Artist's conceptions sometimes de-pict a black hole as a giant funnel in space, but from the outside it looks more like a planet—a big, suspiciously dark planet. Material can and usually does orbit around it. But if you tried to touch what you thought was its surface, your hand would just pass through; the object is empty space. The supposed surface, or "event horizon," is really just a hypo-thetical point of no return, where infalling gas or other matter could reverse course only by exceeding the speed of light. For Sagittarius A°, the event horizon is a sphere about 25 million kilometers across. Matter crossing it just keeps on going, like a car entering a one-way dead-end street, hurtling toward some uncertain and presumably un-happy fate. "It's the one unique feature of a black hole," Narayan says. "The black hole has no surface, and that makes all the difference. The gas and all the energy it's carrying is just swallowed."

Well, what happens to the stuff, then? That's the puzzle. Unfortu-nately, the two main theories that physicists have—gravity theory and quantum theory—reach diametrically opposite conclusions about the fate of swallowed material. Simply put, gravity theory says that falling into a black hole is irreversible, while quantum theory says nothing is irreversible. The former says stuff can't get out; it's assimilated into the black hole permanently. The latter says stuff must get out and partake in the life of the cosmos once again. What gives? This contradiction is a red warning light that some seemingly essential principle of modern physics must be wrong.

Narayan's observations can't settle the issue. Resolving the contradictions of black holes will take a unified theory of physics, one that fuses quantum theory and gravity theory into a quantum theory of gravity. And many of those who have been seeking such a theory think the suspect principle will prove to be locality. If material can travel faster than light or vault from inside to outside without crossing the intervening space, it can slip the hole's surly bonds.

A leading proponent of this idea is Steve Giddings. He is a professor at the University of California, Santa Barbara, although, with his cargo shorts, fleece jacket, and untucked plaid shirts, you might mistake him for a mountain trekking guide. Which isn't too far from the truth: he has appeared in both *Scientific American* and *Climbing* magazines. Giddings is accomplished at rock and ice climbing, downhill and cross-country skiing, mountaineering, and kayaking. He sees his scientific and outdoors passions as complementary. "I feel that these are two sides of intimately relating to nature," he says. On camping trips as a boy, he read physics books; in college, he got a National Science Foundation grant to do research on gravitation, while on weekends he skied the backcountry. The summer after graduating, Giddings built his own kayak and paddled the Colorado River through the Grand Canyon. Then he hitchhiked to Denali National Park, the first of several trips there. He remembers watching a caribou and its calf cross his path, oddly untroubled by his presence. "Then I looked back and I saw why they weren't fazed by me," he says. "They were escaping from a big grizzly bear. At that point, the bear decided to follow *me*." Recalling a park ranger's briefing, Giddings stood his ground and yelled at the bear until it shuffled off in search of easier pickings.

And then he moved to New Jersey. New Jersey has many charms, but mountains and canyons are not among them. Not that Giddings had a life anymore. Days, nights, weekdays, weekends revolved around cramming for exams. Princeton's graduate physics program seemed intent on capsizing his kayak. "There wasn't a lot of encouragement," he says. "It was an atmosphere where the students felt very intimidated." Giddings considered running, but stood his ground, and just as he passed his exams, in 1984, the field of theoretical physics lit up with excitement. Researchers around the world dropped everything else they were doing and took up string theory, a proposed unified theory of nature.

String theory gets its name from the idea that subatomic particles are tiny rubber bands or guitar strings. What we perceive as different types of particles are really just those strings vibrating in different ways, making the world a symphony of unimaginable complexity. The theory had languished in obscurity since the late 1960s, and a tipping point came when its few proponents managed to persuade the majority of its internal consistency. "That was the thing to do—I got swept up in the waves," Giddings recalls. The theory's leading figure, Edward Witten, asked him to solve a crucial formula, and after months of struggle, trying one mathematical technique after another, he nailed it. Meanwhile, he met some fellow kayakers and discovered that the Garden State wasn't completely undeserving of the name. "I started to realize, maybe this could work," he says.

Resolving the contradictions of black holes was one of the main reasons to seek a unified theory, and in 1990 Giddings decided to retrace the steps that led to a paradox, laid out by the celebrated University of Cambridge theorist Stephen Hawking in the mid-'70s. Hawking's starting point was that decay is the rule of nature. Almost everything in this world eventually dies. Black holes are not exempt—nor could they be, if they are to form in the first place. Ruin is creation in reverse. "If you can make a black hole out of random junk, then you can have a black hole decay to random junk," Giddings says.

By Hawking's analysis, decay doesn't mean that the black hole's innards ooze out. How could they? To escape the event horizon, the innards would have to ooze faster than light. Instead, the hole rots from the outside in. The horizon throws the electric field, magnetic field, and other force fields off balance, causing them to shed particles like so many rust flakes. A black hole equal in mass to our Sun releases about one particle per second, which is far too feeble for astrophysicists such as Narayan to detect with their instruments, but enough over trillions of years to reduce the hole to a jumbled, formless spray of particles. The structure the infalling matter had, the information it embodied, every last trace of its identity—all is lost. In other words, the black hole is not just irreversible in the sense that stuff can't get back out. That might not be so troubling, since, from a godlike perspective, you could peer into the hole and reconstruct how objects came to be there. But the hole is also irreversible in that it obliterates matter with such thoroughness that not even a god could recover the original.

As Hawking himself pointed out, his calculations were tough to do. He could track how the black hole affects the outgoing particles, but not how the outgoing particles affect the black hole—and this reciprocal effect could conceivably open a back door between exterior and interior, allowing infalling matter to reemerge. If so, falling into a black hole would be reversible after all, and the paradox would evaporate. So Giddings and several colleagues did a new analysis, based on string theory, to search for escape hatches and hiding spots Hawking's calculations might have missed. They found none. Hawking had been right. "In these simple models, you really vindicate Hawking's original picture," Giddings says.

So, there's no easy way out of the paradox (let alone the hole). Some assumption that goes into the argument must be wrong, and there are only really two assumptions: reversibility and locality. At first, Hawking faulted the former. He suggested that quantum theory is wrong and falling into the hole is irreversible. Yet quantum mechanics seems to be an all-or-nothing package: if it breaks down anywhere, it breaks down everywhere. If it misfires in the way Hawking supposed, we should see parallel failings in ordinary settings, and we don't. Hawking eventually came to agree that black holes must be reversible. By default, locality must be wrong. "I'm continuing to beat my head on the question of how the information gets out—it just seems to have to be nonlocal," Giddings says.

A few others were coming to much the same conclusion. But the general mood was dubious. Nonlocality in black holes is even harder to swallow than nonlocality in particle experiments. Whereas quantum entanglement is subtle and does not overtly contradict any other law of physics, faster-than-light motion across an event horizon is about as unsubtle as you can get. It flouts the Einsteinian speed limit as brazenly as driving 90 mph in front of a state trooper. Giddings could hardly walk down the hall or get a cup of coffee without a colleague challenging his willingness to consider nonlocality, and he ended up dropping the subject for nearly a decade. "It seemed pretty crazy," he says. "I didn't pursue it. I gave in too quickly to the skeptics." But he was only just slightly ahead of his time.

The Cement Mixer and the Sculptors' Daughter

The mere possibility of a second type of nonlocality is hugely significant. It hints that the phenomenon that Einstein identified is one tile in a larger mosaic. That doesn't prove that nonlocality is really operating or that the two types are linked, but psychologically it matters. In science as in life generally, it's usually the second case, rather than the first, that gets people's attention. With a third, you have a trend.

This next type of nonlocality I'll talk about isn't as well established as quantum entanglement or black holes, but if true is more dramatic still. It comes out of observations that seem so obvious you may not even realize they're observations. If you go out to look at the night sky, it's dark. This probably doesn't come as a revelation. Yet the darkness of the night is one of the foundations of the big-bang theory, for it means that the universe is finite in age or size or both. If the universe were infinitely big and old, we would see forever in every direction and our sight line would always intersect a star. The stars would form an unbroken wall of light. It would be like living in a forest so deep and mature that you see a tree wherever you look. So the next time you look at the night sky, imagine the stars as trees and the blackness between them as gaps, revealing that the forest is either so small that you are seeing through to the other side or so young that it hasn't filled in yet.

Not only is the night sky dark, it looks pretty much the same everywhere. At a conference I went to in 1996, astronomers unfurled a poster with the most remarkable demonstration of this uniformity I've ever seen. They had pointed the Hubble Space Telescope at a dark patch of sky near the Big Dipper and kept it focused there for ten days, gathering light for the most sensitive image ever made—the Hubble Deep Field. Three years later, they did the same on nearly the exact opposite part of the sky, in the southern hemisphere. The images are not splashy works of art in the way that some Hubble images are; their beauty is

understated. They show objects at nearly the outermost range of our
vision, so dim that the telescope collected just one photon of their light
per minute. The thousands of little reddish splotches in the image are
entire galaxies, including some of the very first to have formed. The
northern and southern images look statistically the same, which poses
a paradox that the University of Maryland professor Charles Misner
had first realized in 1969.

Misner, a contemporary of Hawking, is another of the physicists
who revolutionized the study of black holes and the universe at large in
the 1960s and '70s. Like most physics students, I first learned his name
as the M in "MTW," the universal abbreviation for the coauthors of the
standard textbook of gravity theory, *Gravitation*—with its vignettes
and musings, among the few science textbooks that is genuinely fun
to read. Misner's childhood interest, like Galvez's, was not physics but
chemistry. He remembers how his mother once complained about the
holes he'd burned in his clothes with his chemistry set. His response:
make an experiment of it, spilling acids on various textiles to see how
they reacted. His mom might not have been entirely pleased, but a
family friend heard of his exploits and hired him to find a way to make
concrete cure more efficiently. He developed a sealer to slow the evaporation of water.

Misner entered college as a chemistry major, and suddenly all the
fun drained out of the subject. "The labs were awful," he recalls. "You
just had to follow the cookbook instructions." So he switched to physics
and went on to grad school at Princeton just as the legendary professor
John Wheeler (W in the textbook moniker) was sparking the study of
gravity back to life. Although physicists professed undying admiration
for Einstein's theory of gravitation, few had actually bothered to study
it, thinking the real action was in quantum physics. Wheeler recognized gravity as the most creative of nature's forces. Geons, spacetime
foam, wormholes, black holes—you don't even need to know what these
terms mean to appreciate that Wheeler was talking about more than
apples falling on people's heads. "He had a geometrical and physical intuition and had this excitement that there might be more in the equations than people had thought about," Misner says. "And he was right."

The uniformity of the night sky didn't register as a mystery until two observational breakthroughs in the 1960s. First, astronomers discovered quasars—points of light that look like stars at first glance, but have colors like no star anyone had ever seen. It dawned on them that quasars look so vivid because the universe is growing, stretching light waves like a logo on a spandex shirt, turning blue to red. Quasars' light is so red that it must have been traveling for billions of years to reach us, making them, at that time, the oldest things human beings had ever seen. In 1966 Misner did a yearlong fellowship at Cambridge, where he remembers astronomers plotting quasar positions on a globe-shaped chalkboard. Misner noticed more chalk marks on one side than the other, as if the ancient universe was lopsided. The skew turned out to be a fluke, but it got him thinking about why the sky should or shouldn't look the same in every direction. A second discovery forced the issue: the cosmic microwave background radiation.

Radio astronomers first noticed this radiation as a faint but constant hiss in their receivers. They scraped the pigeon droppings off of their antenna, and still the hiss persisted; wherever they pointed their antenna, there was this hiss, filling the sky, leaving no empty spots. Researchers quickly recognized the hiss as a form of light stretched from blue to red to infrared to the microwave part of the spectrum, an even more dramatic transformation than quasar light underwent, making its source even older: on current estimates, 13.8 billion years old. The cosmic microwave background radiation gives us a snapshot of the universe at that time, and it looks like a white cow in a snowstorm: an almost featureless primordial soup of nearly pure hydrogen gas. The gas was even more uniformly spread out than the galaxies and quasars that came later.

A common appearance demands a common cause. If two of your friends show up one day wearing exactly the same outfit, you might dismiss the coincidence as pure chance, but if you notice lots of other sartorially coordinated people, too, you assume a connection: a dress code, a mass e-mail, a sale at the local Gap. People can dress in so many ways that it's improbable for them to pick matching clothes at random. Likewise, matter in the early universe could be arranged in so many ways that it was unlikely—breathtakingly unlikely—to have the same density and temperature everywhere. Yet it did.

What could possibly explain this homogeneity? Gravity would act to pull matter together, making it *less* uniform, if anything. Cosmologists have speculated about other processes, but they face a very basic problem. Two galaxies or lumps of primordial gas on opposite sides of our sky, at the outermost range of our vision, are too far apart for any process operating within space to have leveled them out. After all, that's what it means to be at the outermost range of our vision: light from each galaxy is only now reaching us after an odyssey of billions of years. It hasn't had time to cross to the other galaxy yet.

Cosmologists draw an analogy to the horizon on Earth. Because of our planet's curvature, if you stand on a life raft in the middle of the ocean, you can see about three miles out. If two ships approach, one from the north, the other from the south, you first see the tips of their masts, and as they get closer, their hulls slowly rise over your horizon. For their part, sailors on the ships first see the top of your head, and the rest of you gradually comes into view. But at the moment of first sighting, sailors on one ship can't see the other ship—when you are just on their horizon, the other ship is still over it. We are like that castaway, and two diametrically opposite galaxies are like those sailors. We are seeing galaxies that haven't yet seen each other, let alone had time to exchange energy or material that might have homogenized their appearance. The background radiation should be a ragtag quiltwork rather than a seamless glow. "It's extremely difficult to explain why the sky is not extremely mottled . . . ," Misner says. "Observations showed that things were coordinated that never had any physical possibility of communicating with each other."

There's a certain déjà vu all over again about this situation. Far-flung parts of the universe have coordinated their properties in apparent violation of the speed limit set by light. That looks spookily like what Galvez sees in his lab, except that now we're talking about whole galaxies rather than little particles. In 1970 the Russian theorist Yakov Zel'dovich ventured that a type of quantum nonlocality might explain the cosmic uniformity. In general, though, cosmologists were reluctant to go that far. Most took the puzzle as a failing of Einstein's gravity theory and figured that the answer would have to await the unification of physics. In other words, Misner says, "No one thought the Einstein equations could be trusted at those extreme times."

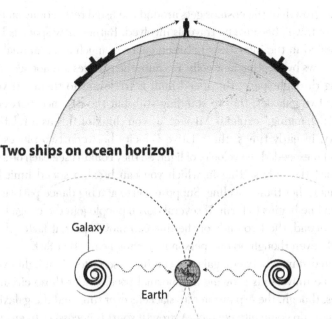

Two ships on ocean horizon

Two galaxies on cosmic horizon

1.3. Cosmological horizon problem. We can see two ships on the ocean horizon even when they can't see each other (*top*). Similarly, we can see two galaxies on our cosmic horizon even when they can't see each other (*bottom*). If those galaxies have never been in contact, what could have made them look so similar? (Illustration by Jen Christensen)

In the late 1970s Russian and American physicists hit upon a way to solve the horizon problem without ditching either locality or Einstein's theory. The idea is that the two galaxies on opposite sides of our sky (or really their precursors) actually *used* to be close together and were dragged apart when the universe went through an early growth spurt. That way, some kind of process would have been able to make them uniform. Like twins separated at birth and brought up not even knowing of each other's existence, the galaxies were once nestled together, but developed independently and only now are becoming reunited again.

For this explanation to work, the growth spurt must have pulled the galaxies apart faster than light, so that they would lose contact with each other until the present day. Normally, the words "faster than light" are physicists' equivalent of scraping your fingernails on a blackboard.

But the growth of the cosmos gets around the usual restrictions on the speed of travel, because no travel is involved. Rather, new space is being created in the interstices between galaxies, much as an animal or plant grows by creating new cells. Because the galaxies are not actually moving through space, the speed limit is irrelevant to them. "If you look at two galaxies, they're standing still, but the distance between them is changing," explains Misner. "If you think of this as a relative velocity, in early times, the relative velocity between two pieces of matter far exceeded the velocity of light. So they couldn't see each other." This isn't the only setting in which you can break a speed limit by growing rather than traveling. Suppose you're at a big dance party and a conga line begins to form. If several dozen people join the conga line every second, the two ends of the line can move apart at faster than 55 mph, even though no one person is moving nearly that fast.

The distance between galaxies can increase faster than light even when the universe is growing at a normal pace. Under those circumstances, though, the expansion rate slackens over time and the galaxies eventually do come into contact. A growth spurt is necessary to ensure that galaxies can have a common birth and then *lose* contact.

Most cosmologists find this concept, known as inflation, so elegant and compelling that they commonly present it as settled fact. In 2014 a team of observers announced they'd discovered the telltale patterns of inflation in the microwave radiation—ripples produced by the mechanism that drove the growth spurt. Commentators were careful to utter the standard disclaimers ("if true"), but clearly took the result as real; they'd expected it for so long. Yet the finding fizzled a few months later, rekindling doubts that even some of inflation's originators have expressed. The main worry is that inflationary theory presupposes the very condition it is supposed to produce: for the universe to begin inflating, it must have already been preternaturally uniform. Consequently, some physicists have been exploring alternatives to inflation, including not just the *appearance* of nonlocality, but the real thing.

•

One of the inflation skeptics is Fotini Markopoulou. I first met her at a conference in honor of Wheeler, where she shared first place in a competition for up-and-coming physicists. I was struck by her thinking about how physics theories need to assimilate that we are part of the universe, rather than outsiders looking in. "One thing that really interests me is the idea that you're inside the universe that you're trying to understand, that you *can* understand it," she tells me. "There is an interesting interplay between being inside the system you're trying to study and also being able to pretend you're not, which is kind of what science is really about." All areas of science feel this tension between inside and outside view, but it is worst in cosmology, the one field that studies a system with no outside at all.

Markopoulou says the big picture has engrossed her from an early age. "As a kid, I would like to go into a church when it was empty and sit down and just look at the ceiling," she recalls. "Greek Orthodox churches basically have—it's like a planetarium. There is this cosmology that is painted on the ceiling. It's something that, for whichever reason, I've always found fascinating. There is something amazing about the idea of a human trying to look at the whole thing that we belong to." It would be easy to draw a straight line between this childhood wonderment and her career as a physicist, but Markopoulou resists that convenient narrative. She also loved art—both her parents were sculptors—and archaeology and architecture. She didn't know what major to list on her college application. When her high school principal suggested theoretical physics, she put down theoretical physics. In college, a friend raved about a lecture series on quantum mechanics, and it happened to be on her way home, so she stopped by. "I did not read a book about Einstein and decide to continue where Einstein left off," she says. "I ended up, at various forks in life, deciding that theoretical physics would be the most interesting choice."

Likewise, she came late to her choice of subject: unifying quantum theory and gravity theory to produce a quantum theory of gravity. As an undergraduate and first-year grad student, she studied particle physics instead. Yet she found her coursework dissatisfying. "It's a funny thing when you're educated as a physicist—you don't really learn quantum theory in a fundamental way," she recalls. Her classmates and professors dismissed unification as a pipe dream, and initially she felt the same. After a while, though, she began thinking that maybe it was all right to dream. Although the answers to cosmic mysteries might lie beyond physicists' grasp, at least quantum-gravity researchers were reaching. "When you asked interesting questions, like, 'Why it is so?' . . . it always seemed like you were not really supposed to ask," she says. "The people who worked on those interesting questions were the quantum-gravity people." Eventually the lure grew too strong to resist. Whereas Giddings has sought to unify physics through string theory, Markopoulou settled into a different community of physicists who take alternative approaches to reconcile gravity with quantum theory. Unlike string theorists, this group does not strictly seek to unite *all* of physics, with its enormous variety of particles and forces, but concentrates on gravity.

Markopoulou made her name by studying whether various proposed quantum theories of gravity obey the principle of locality and showing that most do not. Conventional wisdom has it that such anomalies would be confined to tiny scales, smaller than even an atom, but Markopoulou doubts that anything so profound could be so narrowly contained. "From the beginning, when I started in quantum gravity, I had this gut feeling that maybe the signature of quantum gravity is really large, because you are changing something so fundamental," she says. Markopoulou suspects that the synchronicity of distant galaxies may be that signature. The unity of the cosmos may be a third type of nonlocality writ large. Several string theorists have been thinking along the same lines. "The horizon problem is nonlocality staring us in the face," Markopoulou says.

Particles in the Basement

In the instances of nonlocality I've talked about so far, space is failing in its most basic function: to separate things from one another, to space them out. Entangled particles coordinate their behavior without exchanging signals through space. Matter falls into a black hole and manages to climb out of the abyss of space. Galaxies look alike across an unbridgeable gulf of space. These phenomena give, at the least, the appearance of nonlocality. But as a fourth and final example, I'd like to turn the tables and consider a phenomenon that gives the appearance of *locality*, but might ultimately be nonlocal.

Physicists usually think of the world as made up of particles: electrons, protons, and all the other subatomic creatures of physics. Particles are the very embodiment of locality. These little specks of matter exist at specific locations. They interact with one another only by smacking together or by firing off a middleman particle that shuttles between them. Quantum entanglement can make particles match in a nonlocal way, but doesn't alter this basic picture. Yet it turns out that the entire concept of localized particles is awkward, even self-contradictory.

Lest you think that particles are a remote and abstract notion, they're surprisingly easy to see for yourself. One evening, I went down to my basement with a plastic party tumbler, a foil cupcake liner, a bottle of rubbing alcohol, and one of those spray cans you use to blow crumbs off computer keyboards. Inspired by the simplicity of the experiments I'd seen Galvez perform and one too many episodes of *MacGyver*, I was determined to use these common household items to build a particle detector. A duster can, sprayed for more than a second or two, gets very cold—cold enough to cause alcohol vapor trapped in an inverted cup to condense along the paths of electrically charged particles like tiny contrails.

I worked on that kludgey contraption for a couple of weeks, trying out various designs without success, ultimately combining several ideas to create a device that could hardly be simpler. That's what science is like: hours of frustration punctuated by moments of rapture. When my little device finally worked, I watched as short white streaks betrayed the presence of errant subatomic particles flitting through my house. On occasion, I saw their trails kink sharply, possibly indicating a collision of two particles. My wife was just happy that I hadn't taken apart the clothes washer.

My party tumbler was a miniature version of the giant particle detectors at the Large Hadron Collider. I visited the collider in the summer of 2007 as the machine was nearing completion. I took an elevator forty stories down and entered an underground chamber big enough to fit a cathedral. It was packed with equipment. What awed me the most was not the size of the apparatus, but the sheer number of data cables.

Some 1,800 miles of them flowed through the chamber like a million tributaries of a mighty river. Right at the center was a metal tube barely wide enough to insert a couple of fingers. When the collider is operating, streams of protons surge through this tube like bicyclists in a peloton. Some of them collide, spraying debris all over the underground chamber.

Ever since the late 1940s physicists have visualized particle collisions using a system of stick figures called Feynman diagrams, after their inventor, the renowned Nobelist Richard Feynman. His technique is extremely powerful and extremely accurate. But it is also cruelly difficult. The diagrams look simple, but merely put a brave face on mathematical trench warfare. Zvi Bern, a UCLA physics professor who specializes in these calculations, says he entered grad school enamored of the elegance of Feynman's technique, but soon got a reality check. "I do remember the very first time I got homework in my particle physics class," he says. "I was just amazed that people could actually do these Feynman calculations without making any mistakes. It wasn't even that complicated by the standards of what the pros do, but after twenty pages of algebra I was completely mystified how the pros did it without making mistakes."

The calculations are misery-inducing for two reasons. First, a particle collision has an enormous diversity of potential outcomes. For instance, when two gluon particles—building blocks of the protons that circulate through the LHC—collide, anywhere from two to an infinite number of gluons might fly back out. Second, each of these potential outcomes can be achieved through a huge variety of possible intermediate steps. For instance, two colliding gluons can spawn four gluons in 220 different ways, not even counting the detours they might take in the process. The resulting equations contain tens of thousands of algebraic terms. And that's an easy one. Pity those who consider eight outgoing gluons, for they must account for 10 million possible intermediate steps. Even computers quickly reach their limits.

No one goes into particle physics expecting it to be easy. To the contrary, lots of students are drawn to the subject precisely because it's hard. But if you go through all that effort, you expect to discover something that makes it worthwhile. And you don't. Those tens of thousands of terms eventually reduce to a mere four. The rest cancel one another out. Term number 2,718 might, on examination, be the same as term number 3,142, but with a minus sign in front of it, so you cross off both. Unfortunately, you can't tell in advance which terms will cancel, so you need to write them all out. The procedure seems perversely wasteful, barely better than writing lines in after-school detention. The mismatch between the difficulty of the calculation and the simplicity of the answer suggests that physicists are missing something, like a police captain who rounds up all the usual suspects and overlooks the guy standing there with a pistol in his hand.

Bern's classmates put the homework from hell behind them, but he never got over it. He figured there had to be a better way to do those calculations and threw himself into finding it. It wasn't the smartest career move. Most physicists saw these calculations as journeyman's work: useful but unimaginative. Prospective employers didn't attend his talks; a journal rejected his first paper on the topic as "not very interesting." The breakthrough came when he gave a talk at Princeton and Witten, the renowned string theorist who had brought Giddings back from the brink of quitting physics, came up afterward to compliment him. With this imprimatur, Bern finally got a job. Bern says the whole experience disabused him of his youthful romantic views of science. "Science is not done the way I thought it was," he says. "My epiphany about science is that you have to have luck on your side."

Through his and his coworkers' efforts, physicists no longer need to write out those ten-thousand-plus algebraic terms, but can jump straight to the final four. But why were the old techniques so mismatched, and why do these new techniques work so well? Another theorist, Nima Arkani-Hamed at the Institute for Advanced Study in Princeton, blames nonlocality. Theoretical physicists are known for their strong personalities, but Arkani-Hamed is a force of nature. He was born in Houston in 1972. Several years later, his father, a prominent Iranian geophysicist, moved the family back to Tehran to help build a new country after the shah fell. The family's idealism quickly faded. They criticized the ayatollahs once too often, went into hiding to avoid arrest and likely death, and fled over the Turkish border on horseback.

Lots of physicists will tell you they are "excited" by such-and-such a discovery. They tell you this in such a monotone that you can't help wondering, if that's what they're like when they're excited, how awful it must be when they're bored. Arkani-Hamed, on the other hand, tells you about the simplest things with such verve that you feel he just opened the Lost Ark of the Covenant. Once, he got me to marvel at how the string "1, 2, 3" can be rewritten as "3, 1, 2" or "2, 3, 1," demonstrating how so much of physics boils down to careful counting of possible arrangements. I remember standing with him at a coffee break—and whenever he's around, there does seem to be a lot of coffee being consumed—as the conversation became a rapid-fire internal dialogue, wherein Arkani-Hamed provided his own answers while everyone else was still grasping the questions: "I did this, I tried that, but it didn't work, but—oh, wait, maybe that was because of this—so, hmm, I wonder whether I should . . ."

"I have never been more excited about physics in my life," he gushed when I first asked him about the new calculation techniques. "There is something truly spectacular going on here which I think could eventually change the way we think about both spacetime and quantum mechanics . . . It is developing at a blistering pace right now, with a group of roughly fifteen people in the world working on it day and night." Their effort culminated in 2013 with a comprehensive alternative to Feynman diagrams.

Arkani-Hamed thinks the trouble with Feynman diagrams is that they are expressly local. They show particles as interacting with one another at specific locations in space and time. The diagrams look reassuringly like the trails particles leave in a detector such as the party tumbler in my basement; indeed, this is why physicists were drawn to Feynman's approach. Yet the calculational quagmire brings this feature of the diagrams into disrepute. Locality is directly responsible for the algebraic bloat. "You've insisted that the theory be local," Arkani-Hamed says. "You then suffer through ten thousand terms." By taking every point in space as strictly independent of every other, the Feynman technique overstates the amount of complexity in nature. Much of what appears in the diagrams doesn't exist in the real world, such as "virtual" particles and "ghost" fields. Theorists must impose special rules to ensure that these unwanted guests don't stay for dessert.

Rather than take locality as the starting point, Arkani-Hamed, Bern, and their colleagues assume that particles satisfy certain principles of symmetry, and the resulting equations are much simpler. Particles still obey the principle of locality; the only difference is that the theory derives locality from deeper considerations, instead of presupposing it. This approach combines modesty with enormous ambition. These theorists didn't set out to create a grand new theory of particles, but merely to streamline the existing one. Their equations don't predict anything exotic, but just make it easier to describe what we already knew.

Historically, such reformulations have been hugely significant. This brings out a remarkable fact about theories in physics. They are not fixed structures, but have a kind of ineffable existence that goes beyond any given set of equations physicists use to express them, like a story that can be told and retold with vastly different settings and characters, but is recognizably the same story, or a musical work that can be given a new arrangement, bringing out qualities no one had appreciated before or, conversely, dispensing with inessential flourishes. Perhaps the most dramatic example was when Nicolaus Copernicus put the Sun rather than Earth at the center of the universe. At the time, his model was little more than a mathematical reformulation of the old Earth-centered system, and astronomers adopted it as a streamlined way to draw up calendars and planet charts. But the new conception of the cosmos invited questions that were meaningless in the old. What causes dropped objects to fall? Do planets' orbits have to be circular or could they be oblong? Could space go on forever? Copernicus's work may not have been revolutionary, but it fomented revolution.

So you see, laziness has its benefits. People trying to get away with doing less work are a force for innovation. Arkani-Hamed hopes the reformulated particle theory will unstick the search for a unified theory of physics. Once you stop assuming the world revolves around locality, the pieces begin to fall into place.

What principle should replace locality? If the world doesn't really consist of localized particles, what *does* it consist of ? No one yet knows. But now physicists have a path forward. Whereas Einstein agonized that nonlocality would mean the collapse of contemporary physics, Arkani-Hamed thinks it marks the rebirth. "When you're a kid, this is what you fantasize about what it's like to do theoretical physics," speaks the man who never really stopped being a kid.

·

We've seen nonlocality pop up all over the place: in experiments on the quantum realm, in the paradoxes of black holes, in the grand structure of the universe, in the maelstrom of particle collisions. In all these examples, physics enters a twilight zone. Things can outrun light; cause and effect can be reversed; distance can lose meaning; two objects may actually be one. The universe becomes spooky.

Although these varieties of nonlocality come up in different contexts, they have some striking commonalities that suggest physicists are feeling different parts of the same elephant. Arkani-Hamed, for example, thinks the type of nonlocality in his theory may subsume quantum entanglement. "It's not inconceivable that proper understanding of these things will lead to a new way of talking about quantum mechanics, not just spacetime," he tells me. "Maybe in this new picture there might be some new picture for what entanglement means." It goes the other way, too. Giddings and others think quantum entanglement might be the glue holding space together. Entangled links might even create a kind of secret tunnel between the inside and outside of a black hole. The coming chapters will explore these fascinating ideas.

Blown minds are an occupational hazard in physics. This is a profession whose entire purpose is to look beneath appearances for a world that's simpler than it seems, yet also more remote from our experience. What's interesting, though, is that this isn't the first time that physicists and philosophers have encountered such mysteries. In many ways, the history of locality is the history of physics itself.

2

The Origins of Nonlocality

What's the big deal about nonlocality? Why can't scientists just throw locality into the same dustbin as phlogiston, vortex atoms, and all the other beautiful hypotheses slain by ugly facts? Why does nonlocality inspire such melodramatic putdowns: "the end of the rationality of physics," "incompatible with the very possibility of doing science," "flapdoodle"? Clearly, the violation of locality is not just a routine occurrence in the churn of ideas, something you can laugh off over a beer after work. To see why, you need to dive into the history of physics, for nonlocality threatens the very nature of what we assume physics to be.

Physics is not like other sciences. If you ask geologists, biologists, or astronomers to define their subject, they can point to rocks, things that slither, or twinkles in the night sky. Physicists, however, start pointing at everything around them; they're not particular. You are as apt to see them studying the origami of biological proteins and the yo-yoing of financial markets as the collision of tiny particles. Their discipline is defined less by its subject matter than by its goals. In whatever they focus on, physicists seek the simplicity in complexity and the unity in diversity. Like philosophers, their intellectual siblings, they are driven by the conviction that the universe is within the human power to understand and that if you look beneath its variety and intricacy, you will find comprehensible rules.

Also like philosophers, physicists look to history for guidance for what those rules—and, by implication, their discipline—should be. They have a reputation as some of the most forward-looking of all scientists, so far ahead of the technological curve that they *create* the curve. Physicists can rightfully claim credit for almost every gadget you own. Yet I find that they spend as much time looking backward as forward. They routinely cite centuries-old developments or tell me they're reading a biography of some illustrious figure, on the assumption that you can't go forward until you know how you got where you are.

And indeed, physicists' general standards for simplicity and comprehensibility have remained remarkably constant over the ages. Their intellectual forefathers in ancient Greece sought to describe the universe as a giant billiards game. The balls—the building blocks of the world—fly around, smack into one another, and rebound in an endless chain reaction. These interactions are strictly local: until balls touch, they exert no effect on one another. Though individually simple, the balls and collisions are so numerous that they give rise to all the rich diversity and complexity of the world. To a degree its inventors really had no right to expect, this picture captures something of reality. Through the coming millennia, the details changed thoroughly, but the core values endured. Especially locality.

Sure, even the ancient Greeks were well aware of counterexamples to this principle. They didn't know about quantum particles or black holes yet, but they did know about other effects that appear to act nonlocally, notably the phenomena we now associate with the force of gravity. But they didn't give those counterexamples much weight. Most reckoned that the apparent instances of nonlocality were merely a false impression, awaiting some smart person to explain them by processes operating locally. To give up locality would have been tantamount to giving up physics.

What's puzzling, though, is that Isaac Newton did exactly that. To explain those gravitational phenomena, he gave up physics, at least as it was defined at the time. For that, he is remembered as the greatest figure of the Scientific Revolution, the intellectual ferment in the seventeenth century out of which science as we know it today emerged. And the reaction that Newton's theory of gravity originally provoked has remarkable parallels to the dismay I hear about non-locality today.

•

Frans van Lunteren, a historian of science at Leiden Observatory in Holland, one of the most storied scientific institutions in Europe, recalls how unsettled he felt when he first learned about Newton's law of gravity. His high school teacher explained that apples fall and planets cling to the Sun because everything in the universe exerts a force of attraction on everything else. As Newton conceived of it, the force acts at a distance instantaneously. Raise a finger on Earth and all the distant planets in the universe instantly quiver (a little). Gravity leaps from Earth to apple, and from finger to planets, without passing through the space between them.

This is what weirded out van Lunteren as a teenager. "I found it hard to understand how a lump of brute matter—say, a piece of rock—could affect some other matter deep down in space, especially when the intermediate space is vacuous," he says. But he figured that if he didn't understand it, the fault must be his. "I was quite used to missing the point of something," he confesses. Only as an adult did he learn that this strange feature of gravity goes by the name of nonlocality.

At the time, van Lunteren didn't care for history. He dropped his history classes in order to focus on math and physics. But when he got to college, his physics classes were a letdown. They consisted of equations and more equations, with nary a word for what all those x's and y's actually meant. "Most teachers would start with a differential equation on the upper left side of the blackboard and then work themselves toward a cross section or some other measurable quantity on the right down side, without having anything interesting to say about the physics," he recalls. He blew off his classes, read French and Russian novels instead, worked odd jobs, hitchhiked to Istanbul. In his intellectual wanderings, he learned about the strange quantum phenomena that so troubled Einstein. When he recovered his passion for college life, he found himself pulled in by history. He saw it as a way to explore grand intellectual questions his physics professors had ignored. Like the philosopher Tim Maudlin, van Lunteren found that to love physics, he had to leave physics.

Casting around for a Ph.D. topic, he went back to his teenage puzzlement over the nonlocality of Newtonian gravity. He discovered he had been right to be puzzled; those who accepted the theory uncritically had missed the point. Nonlocality troubled Newton himself as well as his peers. It sounded pseudoscientific, like astrology or miracle cures. One French mathematician whined, "We are plunged back into the old darkness." Van Lunteren says, "It would have helped me if my high school teacher would have added that many great contemporaries of Newton found that notion hard to swallow or even incomprehensible." So he wrote his dissertation on how scholars did their utmost to explain away Newton's action at a distance.

In the end, scholars were not plunged into darkness. They pivoted. The generation that grew up with Newtonian gravity found the theory entirely natural. For millennia, natural philosophers recoiled from nonlocality; in the eighteenth century, they embraced it. Put simply, they were for locality until they were against it. And no sooner did scholars get used to Newtonian nonlocality when along came another U-turn and a new generation went back to thinking that the world had to be—just had to be—local, thereby setting up our present predicament.

The Mechanical Universe

These historical twists and turns began with one of the most famous encounters in Western intellectual history, an event it'd be great to go back and witness if you had a time machine. As Plato tells the story, in the year 451 or 450 B.C., Parmenides, the leading philosopher of the day, and his best-known student, Zeno, traveled to Athens from their native city of Elea in southern Italy. They stayed at the house of a leading politician just outside the city walls. One day, who should stop by but an up-and-coming young Athenian philosopher, Socrates.

The very concept of philosophy (in both Greece and its other birthplace, China) was scarcely a few generations old. It was a radical new way to fathom what happens in nature. In daily life, when we ask "Why?" what we're usually after is what drove a person to do what they did. Traditional mythology extended this style of thinking to the natural world. Why did the earthquake occur? Because Poseidon was peeved that someone had defiled his temple. Such explanations make little distinction between locality and nonlocality. Sometimes gods act nonlocally (they'd snap their fingers and make things happen) and sometimes they act locally (sending an emissary to do their bidding). As far as mythology is concerned, that's a minor detail.

Philosophers were people who found these character-driven narratives unsatisfying. Even if you grant the existence of Poseidon, how could he bring about an earthquake? What rules governed what he was capable of? Philosophers didn't care about the motive; they wanted to know the mechanism. The categories of locality and nonlocality assumed a new importance. Naturalistic explanations tend to be local. In our experience, when you want something to move, you can't will it to do so; you need to go over and push on it or send someone to do it for you. The first philosopher we know by name, Thales, suggested that earthquakes occur because the land is floating on a subterranean ocean like an unstable boat, occasionally rocking back and forth. The cause is in direct contact with the effect.

But locality made Parmenides queasy. He wasn't so sure we could trust our everyday experience, or that we could divide up the world and understand it piece by piece. Defending this thesis to Socrates in Athens, Zeno argued that local concepts such as motion, change, and individuality lead to logical paradoxes. History records nine such paradoxes; dozens more may have been lost to time. The deepest and most influential was that of complete divisibility. If an object could be cut in half, then fourths, then eighths, and so forth without end, it would consist ultimately of geometric points, each a nothingth of an inch across. When you went to reassemble the object, you'd run into trouble because no number of nothingths add up to a somethingth. From this Zeno concluded that reality could not, in fact, be divided up.

Socrates complained that it was all a bit over his head. Zeno's arguments "deny the most self-evident thing," as a later Greek philosopher wrote. But that's what made them so vexing. To claim that nothing consists of smaller pieces sounded crazy, yet the reasoning looked solid. In that house in Athens, Parmenides and Zeno created an intellectual crisis. For decades after, people would travel halfway across the Greek world to hear firsthand accounts of the debate that ensued.

Modern mathematicians think Zeno was basically right that something gets lost when you divide a continuous object into infinitesimal pieces. The number of geometric points in a continuum is uncountable—literally uncountable. And if you can't count them out, you can't add them up. Our usual intuition that the whole is the sum of its parts doesn't apply. The continuum has no innate scale; the size of a set of points does not derive from the size of the individual points but must be defined separately. "One reading of Zeno's paradoxes is that you can't get physical scale at all from a continuum," says the theoretical physicist Fay Dowker of Imperial College London.

Although physicists have made their peace with the continuum, many still find the idea unsettling. The great physicist Richard Feynman wrote: "It always bothers me that, according to the laws as we understand them today, it takes a computing machine an infinite number of logical operations to figure out what goes on in no matter how tiny a region of space, and no matter how tiny a region of time. How can all that be going on in that tiny space? Why should it take an infinite amount of logic to figure out what one tiny piece of space/time is going to do?"

·

Worries of this sort led many Greek philosophers to propose that matter is not infinitely divisible, but composed of discrete building blocks. The atomists were prescient. When you read what has survived of their writings, you feel like you're reading a freshman physics textbook in verse form. Sticklers might sniff that the atoms of antiquity were nothing like the modern variety, but the general conception of nature developed by Democritus and others in the fifth century B.C. was remarkably close to the one that physicists nowadays have. Everything that happens in nature, the atomists asserted, derives from the shape, motion, and spatial arrangement of tiny building blocks. All the sensations we enjoy—taste, color, smell—they believed are produced by streams of atoms that objects spew out and that impinge upon our bodies. The sight of objects literally pokes you in the eye; a bitter taste stabs you in the tongue.

The concept of space was the atomists' creation. They were the first philosophers to argue that matter needs a venue in which to exist and move. One of Democritus's successors, Lucretius, wrote: "If there were no place or space, that which we call / Void, then particles would not have anywhere at all / In which to be or move." Space defines the position, velocity, size, and shape of atoms. It extends forever in every direction, populated with an innumerable variety of worlds. This cosmological picture, radical in its time, proved decisive in atomism's eventual triumph.

If atoms were the athletes and space the playing field, locality was the rulebook. Like modern physicists, the atomists distinguished two aspects of locality. First, space separates atoms from one another and gives each an individual identity. This is the principle of separability that Einstein considered essential to physics and that quantum physics appears to violate. Second, space dictates how atoms affect one another. The atomists held that atoms interact only by making direct contact. Until atoms bang together, they glide through space in straight lines, oblivious to one another's presence. This is an early version of the principle of local action, which Einstein formalized in his theory of relativity. It lets you explain any event as the outcome of earlier events.

The atomists didn't give any real arguments for locality. They didn't even put it forward as a tentative hypothesis to be confirmed by experiments; they didn't have the concept of empirical science yet. Rather, they took locality as a self-evident truth. For objects to act on one another at a distance would break the causal chain of events. It would render the universe incomprehensible.

•

Atomism was the first "theory of everything." A few blind spots notwithstanding, there was hardly a phenomenon of life, the weather, or the heavens for which the atomists did not offer an explanation. They pioneered the mechanical view of nature, a clockwork universe. Modern terms such as "quantum *mechanics*" reflect this heritage. To be sure, Democritus himself didn't think in terms of machines; that analogy came centuries later as machines became more commonplace. When philosophers and scientists speak of a mechanism, they mean simply a system of interlocking parts, as opposed to a contraption designed for some purpose. What atoms do gives them purpose, not the other way around. Individual atoms are lifeless, lacking in willpower and animation. If one moves, it is only because another caused it to. This void of purpose and meaning put off most of Democritus's contemporaries. Plato wanted to burn his books. To this day, physics strikes many people—even physicists—as cold, abstract, inhuman.

Maybe it is. But it is also liberating. Atomism transcended human experience. The old mythological explanations attributed earthquakes to emotions—one complex phenomenon to another complex phenomenon—and what kind of an explanation is that, really? It's just passing the buck. Real explanation means breaking something down into simpler pieces and showing how they interact to produce that thing. Who would go back to the *telenovela* of Greek mythology, in which cities fall or famine sweeps the land because Zeus can't keep his pants on? As the literary critic Stephen Greenblatt described in his Pulitzer Prize–winning book *Swerve*, Democritus's successors created an entire atheist, live-for-today philosophy in which humans create our own meaning. Lucretius wrote: "Nature, rid of harsh taskmasters, all at once is free. / And everything she does, does on her own, so that gods play / No part."

The most famous philosopher of the ancient world split the difference between the atomists and their detractors. From what Aristotle could see, the world throbs with life, and life is purposeful, so it stands to reason that inanimate objects, too, are goal-oriented. An apple falls toward the center of the Earth because that's where it belongs in the grand scheme of things. Its motion is spontaneous, requiring no external cause. Aristotle also went back to the idea that stars and planets are governed by different laws than apples and arrows. And he rejected the atomists' claim that objects are made up of indivisible parts. Zeno's paradoxes notwithstanding, he thought that matter is continuous and developed a sophisticated theory of the continuum that anticipated modern mathematics. Objects' properties cannot be reduced to the arrangement of atoms.

Aristotle abhorred a void. Objects in his scheme fit together like jigsaw pieces without any empty gaps, and a given object's location is defined in relation to its neighbors, rather than by some abstract framework that exists independently of matter. Because even "empty" space is already jam-packed with stuff, light can't be a stream of atoms that cross space from a bright object to our eye. Instead, Aristotle thought of light as an impulse conveyed through a medium. A bright light energizes the medium immediately around it, and a wave of transformation sweeps across space in a continuous movement like a disturbance rippling across a pond. No particles are moving; rather, each little piece of the medium passes the impulse to the next, like kids playing a game of telephone. Aristotle's contemporaries in China also conceived of the world as a continuous medium, *chhi*.

If anything, Aristotle's way of thinking hewed more closely to observation than atomism did. That said, Aristotle didn't attempt to make specific predictions that might corroborate or disprove his theory. Like Democritus, he was concerned above all with making the universe comprehensible.

Despite its organic features, Aristotle's theory adopted many of the essential features of atomism, including locality. It treated nature as a system of objects interacting purely by contact. For an object to deviate from its natural course, something has to push it. Aristotle wrote: "The immediate agent of bodily change of place must be either in contact with or continuous with the moved object, as we always observe to be the case." Also like the atomists, Aristotle took pains to develop a theory of space. To him, having a location was almost the very definition of existence; lacking one, of nonexistence. He wrote: "That which does not exist is nowhere. Where, for example, is a goat-stag or a sphinx?"

Although Aristotle wrote that locality was "universally" true, he mentioned several counterexamples. These anomalies went back to Thales, who remarked on one of our planet's many strange rocks, lodestone, with its ability to tug on chunks of iron. A region in northern Greece known as Magnesia had large deposits of the stuff, giving us the name by which we know such materials today: magnets. Thales also marveled at amber, a piece of which, after being rubbed vigorously with a cloth, will make your hair stand on end. The Greek word for amber is *elektron*, source of the word "electric." Chinese scholars discovered these phenomena at roughly the same time, although they were quicker to make practical use of magnetism than their counterparts in the West were.

It baffled the Greeks how lodestone and amber could influence objects they were not touching. Worse, the influence they exerted was to pull. In a world of contact action, things interact in only one way: they bang into one another and rebound like billiard balls. They push; they do not pull. Trying to explain how push comes to pull taxed the inventiveness of philosophers. Atomists thought these materials released vapors that displaced the air around them, creating a region of low density into which the surrounding air rushed, sweeping up iron or hair with it. Aristotle dealt with the problem by the time-honored strategy of ignoring it.

Magnetism and static electricity were not the only head-scratchers. There were also the phenomena that today we recognize as caused by gravitational attraction, such as falling, ocean tides, and planetary orbits. Aristotle saw them as unconnected. To him, falling is simply objects' wont, tides are generated by winds stirred by solar heating, and planets ride on giant rotating crystalline spheres. The atomists tied these phenomena together and attributed them to the structure of the solar system. In their view, currents of particles surge and swirl through the cosmos, creating vortices in which planets coalesce like a pile of leaves caught in a river eddy. If an object is not keeping up with the rotary flow, surrounding particles push it inward. To "fall" is to be swept toward the middle of an eddy. In short, gravity is not a force of attraction, as scientists later came to think; it's a direct physical push, a shove from above.

•

Aristotle's theory carried weight, literally: in English translation, the surviving writings fill six thousand pages. Scholars in the Greco-Roman empires and the Islamic world built on his work, but much of it was lost or forgotten amid the general decay of European intellectual life in the early Middle Ages. The next major advances in the notion of locality did not come for another two millennia. European scribes began to rediscover Aristotle in the twelfth century, by the roundabout route of

Latin translations from Arabic, and the knowledge so outshone theirs that it must have seemed like an encyclopedia left behind on Earth by aliens from a superior civilization. They transcribed and translated those six thousand pages and spent centuries analyzing them, critiquing them, and harmonizing them with Christian beliefs—a program known as scholasticism. Whatever interested Aristotle interested them. Aristotle thought space was important, so they thought space was important. Aristotle adhered to the principle of locality, so they adhered to the principle of locality. They figured that not even God could evade locality, although the principle was moot in his case: he existed everywhere, so he was automatically in direct contact with everything. "No action of an agent, however powerful it may be, acts at a distance, except through a medium," wrote the leading scholastic philosopher, Thomas Aquinas. "But it belongs to the great power of God that He acts immediately in all things. Hence nothing is distant from Him."

The more scholars mulled Aristotle's theory, however, the more they felt let down, and they broadened their efforts from recovering Aristotle's ideas to improving on them. Aristotle's non-explanations for magnetism and static electricity were a conspicuous weak point. In the late 1500s, the English physician William Gilbert (who later served as Queen Elizabeth I's personal medic) showed that a lodestone attracts an iron ingot even when you put something in between them to block any putative vapors or medium. It seemed unavoidable that magnets acted at a distance. Gilbert was at a loss for a natural explanation and leaned toward a supernatural one: that a magnet is "like a living creature" and attracts iron in an act of "coition."

Aristotelian cosmology also struck many as suspect. How could the universe be finite in size and bounded by a giant rotating crystalline sphere? Such a sphere would have no external reference point to define its rotation. In the early 1500s, this inconsistency was what inspired Nicolaus Copernicus to place the Sun rather than Earth at the center of the solar system. With his cosmic switcheroo, the whole Aristotelian system began to unravel. Aristotle said things fall down because that's the direction of the center of the universe. That is no longer true in the Sun-centered cosmology. So Copernicus created the impetus for an alternative account of gravity. And because the center of the universe no longer defined objects' motion, the universe might as well have no center. It could be infinite—a key tenet of atomism. Having also rediscovered Lucretius's writings, many scholars put two and two together: the structure of the cosmos was evidence for atoms. It wouldn't be the last time that philosophers and physicists learned about the small by looking to the big.

Atomism came into full flower with René Descartes in the mid-seventeenth century. Today Descartes is remembered for the proposition "I think, therefore I am" and for the Cartesian coordinates used on graph paper. But these were just two elements of an ambitious project: to out-Aristotle Aristotle. He wrote to a friend, "I have decided to explain all the phenomena of nature, that is to say, the whole of physics." And he succeeded: his was the first new theory of everything in two thousand years that could claim to be as comprehensive as Aristotle's. Descartes fully integrated Copernican cosmology with mechanical philosophy, and his ideas served as the manifesto of the Scientific Revolution.

Descartes stressed the differences between his theory and classical atomism, perhaps to bolster his claim to originality, but the continuities are clear. Nature consists of particles interacting spatially. An object has no inscrutable innate properties or tendency to seek out its rightful place in the universe, as Aristotle had speculated. It's a purely geometric figure. It has a size and a shape, but no color, no texture, no mass. With just a few numbers (his Cartesian coordinates) to specify an object's position, you know everything there is to know about it. Things could hardly be simpler.

Descartes' objective was comprehensibility—to make the workings of nature completely transparent. Locality was essential to this goal. Objects interacted strictly locally: they moved freely in straight lines until two collided; only then did they change course. Like Democritus and Aristotle, Descartes offered no real evidence for this principle. "These things require no proof, because they are obvious in themselves," he wrote. In everyday life, we must touch objects to make them do something, and Descartes assumed that contact action governs everything else in the universe, too. Trouble was, it doesn't. Descartes did such a thorough job of applying the principle of locality that he inadvertently showed the extent of its failure.

For instance, Descartes espoused the old atomist account of gravitation as a shove from above. In his theory, planets lie at the centers of cosmic vortices, whose swirling motions push particles inward. As an explanation for why planets cohere, there was a lot of truth in his picture. It got the overall shape of the solar system about right and inspired modern theories of planet formation. But the theory stumbled over details. Among its many failings, it implied that objects should fall toward Earth's rotation axis, where the swirling motions vanish, rather than toward its geometric center. If that were true, an apple dropped near the North Pole would "fall" laterally rather than straight down. As for magnetism and static electricity, Descartes attributed them to particles shaped like tiny screws or hooks. The best you can say for this idea is that it wins points for imagination.

Was the mechanical vision of the universe basically OK, in need of just some bolt-tightening? Or did it have to be junked? This is a dilemma scientists wrestle with whenever they encounter counterexamples to a theory. Reasonable people will disagree, and the answer is obvious only in retrospect, and often not even then. In this case, much more was at stake than one theory. To defy mechanical theory and its central assumption of locality was to defy science itself. Would admitting its failure mean that nature was beyond our rational understanding? In a way, the surprising answer was yes. To fix mechanical theory, participants in the Scientific Revolution needed to reach beyond the limits of science itself: to the realm of magic.

Magic in the Machine

If there's one thing you learn in school about science, it is that science is the opposite of magic. From ancient Greece onward, philosophers and scientists have taken on the job of disabusing people of all the crazy things they believe in: homeopathy, horoscopes, and hoodoo, just to take the h's. But then what do you make of the fact that a lot of the greatest scientists in history spent a lot of time trying to do magic? Newton set up an alchemy lab in a garden shed and assembled one of the world's greatest libraries on the topic.

I've never done a rigorous survey, but a large fraction of the scientists

I know have been engrossed by the paranormal at some point in their lives. I myself went through a period in college when I read every book I could find on out-of-body experiences, and I later learned that none other than Richard Feynman had been fascinated by them, too. At a distinguished institution that shall remain unnamed, I joined a study group on alien abductions organized by postdocs and grad students. A number of abductees came to campus to tell us their stories. If anything, we *wanted* to be persuaded, although in the end most of us weren't.

Well, nobody's perfect—that's the usual way scientists interpret this interest. But in the mid-twentieth century, historians of science came to appreciate that magical ideas have been too pervasive to be written off as youthful flings or late-career dementia. Some have gone so far as to say that modern science is as much a product of magic as of mechanism. These represent the two emotional impulses that drive researchers: to delight in the bizarre, and to cut through the BS.

Most of us associate magic with wands and bullying potions teachers, but it's actually an entire mystical belief system. In Western culture, the most influential such systems—Neoplatonism, Hermeticism, and Gnosticism—arose in the second and third centuries A.D. as a backlash against orthodox Greco-Roman religion, which struck many people as too much head and not enough heart. These systems, which mingled with early Christianity and the Jewish mystical tradition of Kabbalah, harkened back to older ideas such as Parmenides' vision of the oneness of nature and Plato's revulsion against mechanical explanations. They remain influential today.

Believers recoiled from the central tenet of mechanical philosophy: that the universe is ultimately simple and comprehensible. They regarded the cosmos not as a clockwork made of lifeless parts, but as an organic unity beyond our power to comprehend rationally. Beneath the reality we observe lies a hidden, or "occult," level that is neither simple nor comprehensible. Objects possess inscrutable properties and powers that tap into this deeper level and can be exploited by concocting charms and potions. In one of Renaissance Europe's leading how-to manuals about magic, Cornelius Agrippa explained: "They are called occult qualities, because their causes lie hid, and man's intellect cannot in any way reach, and find them out." (Agrippa's name should sound familiar to Harry Potter fans: he made a cameo appearance on a chocolate frog trading card.)

Nonlocality was an essential part of these beliefs. A cat's cradle of connections ties together things that might seem unrelated. Small things can affect large ones; what lies here can affect what lies there. These nonlocal influences operate somewhat like human emotions: parts of the universe literally like or dislike one another, creating an all-pervading network of sympathies and antipathies. One of the first anthropologists to study magical beliefs, James Frazer, described in 1911 the two basic principles: "first, that like produces like, or that an effect resembles its cause; and, second, that things which have once been in contact with each other continue to act on each other at a distance after the physical contact has been severed." These influences sound like the very definition of supernatural, but Agrippa, for one, saw them as entirely natural—our fleeting glimpses of the underlying unity of the physical world. As evidence, he and other Renaissance adherents cited those phenomena that had given mechanical philosophers such trouble, including magnetism and ocean tides, along with other ideas that almost everyone at the time (including mainstream philosophers) believed in, including alchemy, astrology, and numerology.

Magical ideas sound dreamy to modern scientists, but, frankly, so do many of the models put forward by mechanists such as Democritus and Descartes. In both cases, what was important in the long term was the general mental framework. Mechanical models emphasize nature's transparency, magical ones its mystery. Mechanical models are reductionist, magical ones holistic. Historically, Western culture has seesawed back and forth between these complementary points of view. The magical perspective enchants people with the thrill of the forbidden, loses its appeal as people start to wonder what they've been smoking, regains it as rationalists get cocky and claim they can purge the world of mystery, and so it continues even today. This cycle has been a driving force in scientific revolutions, and some historians see traces of it in the Bohr-Einstein debates, with Einstein arguing for the rational comprehensibility of the universe and Bohr for its ultimate incomprehensibility.

•

One of the periodic revivals in magical thinking occurred in the fifteenth century. Having transcribed and translated ancient philosophical writings by Aristotle, Lucretius, and others, scholars turned to ancient Neoplatonist and Hermetic texts. Interpreters such as Agrippa latched on to them as a refreshing break from the fussy old scholastic ways.

Two features of magic, in particular, filled in the blind spots of mechanical philosophy. First, it was empirical. At the time, experiment did not have a big place in mainstream philosophy. Both the old-school scholastics and reformers such as Descartes thought they could solve the mysteries of the universe just by thinking hard. Believers in magic, on the other hand, thought nature defied reason. To probe its mysteries, you had to buy some test tubes and get cracking. Practitioners of magic sought not just to study but also to manipulate nature, so as to make the world a better place. They were responsible for much of the idealism of the Renaissance. In the "Oration on the Dignity of Man," Giovanni Pico della Mirandola, a twentysomething Hermetic and Kabbalist philosopher, argued that humans' status comes not from our position in the cosmic scheme of things, but from what we decide to make of ourselves. For that noble sentiment, the pope had him arrested. Among those he inspired was Shakespeare. The *Hamlet* soliloquy "What a piece of work is a man" uses strikingly similar language, and a utopian magician is the protagonist of *The Tempest*.

Would-be wizards failed utterly in the goals they set for themselves. Alchemists could not turn lead to gold, astrologers could not divine the fate of kings, and potion makers could not cure the spleen with green lizard urine. But as John Lennon sang, life is what happens while you're busy making other plans. Alchemists and astrologers developed advanced experimental techniques (hermetic seals, for example) and amassed reams of data, laying the foundations of modern chemistry, medicine, and astronomy. The empirical, idealistic bent of magic provided a model for the pioneers of modern experimental science, who, like scientists of all eras, fancied themselves as rebels. The Englishman Francis Bacon wrote: "The aim of magic is to recall natural philosophy from the vanity of speculations to the importance of experiments."

The second benefit of magic was to nudge philosophers to think outside the mechanical box. It suggested ways for objects to interact besides crashing together—namely, nonlocal ways. The concept of gravity descends from the magical notion of sympathies: objects fall because they are seeking out others of their kind. Earth coheres because stone attracts stone. These magical antecedents are evident in the work of the astronomer Johannes Kepler in the early 1600s. His books read like blog posts. They fess up about false leads, leaps of faith, and crises of self-doubt. He was never one for academic formality: "Woe to me, here I blundered" is more like it. And he was forthright about his mystical inspiration. Kepler cast horoscopes for a living, and, though dubious that he or any other astrologer could predict specific events, took it for granted that heavenly motions guided earthly happenings. He argued that if the Moon were watery, as people at the time generally thought, it would naturally tug on Earth's oceans and raise the tides. Magnetism, which was also taken to be a magical force, could fine-tune planet orbits.

Mechanical purists were unsympathetic. Galileo Galilei thought Kepler had gone over to the dark side, lending "his assent to the moon's dominion over the waters, to occult properties, and to such puerilities." Kepler's ideas orbited on the fringes of mainstream philosophy for half a century, until Newton appreciated how right he had been.

•

One historian has called Newton "the great amphibian." He was both mechanist and magician. Like many of his contemporaries, especially in England, Newton was broadly persuaded by Descartes' mechanical theory, yet disillusioned by its failings. As if its troubles explaining celestial motions weren't enough, the Cartesian clockwork was barely one tick away from atheism. What need was there for God if the wonders of nature reduced to the mindless grinding of gears and springs? Although Descartes had written a role for God into his model, it was rather lame—plainly just a fig leaf to avoid a knock on his door from the pope's agents. Other prominent atomists, notably Thomas Hobbes, gave up the pretense and came out as flaming atheists. For his English compatriots, this was beyond the pale. They were committed to religion, both intellectually and as a matter of self-preservation.

To bring atomism into harmony with religion, Newton and other English philosophers in the mid-1600s fused it with ideas from alchemy, Neoplatonism, and Kabbalah. They thought particles could be animated by "active principles" or a "subtle spirit," as Newton put it. In practical terms, particles could exert and respond to forces that act nonlocally. Forces gave the universe some divine sparkle; though not literally spirits, they bore witness to God's design.

So, if gravity seems magical, that's because it *is* magical. The theory that Newton published in 1687 in his magnum opus, the *Principia*, was still mostly mechanical: the world consists of moving particles obeying strict laws. But it assimilated the magical idea that those particles are linked by a web of nonlocal forces. Newton's concept of gravitation differed from its magical precursors in that it is universal: it isn't limited to objects that have an obvious affinity for one another—stone for stone, water for water—but draws together anything that has mass. It also differed from orthodox mechanical models in that mass is not a geometric property but something lying beyond the scope of reductionist explanation.

For historians, this saga is a case study of how treacherous it is to draw a line between science and not-science. Gilbert, Copernicus, Bacon, Kepler, and Newton, like scientists of all eras, were intellectual magpies, weaving their theories out of whatever scraps they could find. The more flamboyant the scraps, the more original the nest. You can tell creative scholars by their eclectic views. Or as one theorist put it to me: "All good physicists intellectually sleep around." Not that you would know it from reading most scientific books and papers. Like teenagers who swear that their parents never did anything for them, scientists have a way of co-opting ideas from other sources and then denying they had done any such thing. Magic? What magic? Who said anything about magic? But Newton's contemporaries knew perfectly well where his ideas had come from—and the followers of Descartes weren't about to take it lying down.

The Gravity Wars

On March 7, 1693, the renowned German philosopher Gottfried Leibniz wrote a letter to Newton congratulating him on his new theory of gravity, which neatly explained all those phenomena that atomists had struggled to account for: falling, tides, planetary motion. There was no knocking its empirical success. But Leibniz wanted to know what explained gravity. Following in the tradition of Democritus and Descartes, he thought that the nonlocality of gravity had to be an illusion. If you looked closely enough, you should see some local mechanism that causes dropped objects to fall and planets to trundle around the Sun. How else could the world make sense?

Leibniz wrote letters as we write e-mails. Over his life he sent fifteen thousand letters to eleven hundred people. To this day, they have yet to be fully cataloged. And these weren't tossed-off one-liners; many were extended essays that broke open whole new areas of science and mathematics. Like today's frazzled e-mailers, Leibniz complained about information overload. "I cannot tell you how extraordinarily distracted and spread out I am," he wrote to a friend.

Leibniz never met Newton, but for several decades he and his compatriots conducted a debate-by-post with Newton and his. It culminated in five rounds of letters between Leibniz and the English philosopher Samuel Clarke, an exchange that ended only when Leibniz died in 1716. By then the initial cordiality had degenerated into a flame war. The correspondence is rich in ideas, but when I read it I'm struck by how little Leibniz and Clarke really engaged with each other; each kept asserting and reasserting his position, never giving his opponent the benefit of the doubt. To be fair, disagreements over such a foundational question as the nature of space couldn't have been resolved with a smile and a handshake, because people didn't even agree on what would qualify as a satisfying resolution.

For Leibniz and other critics of Newtonian theory, such a resolution had to involve a mechanical explanation. By failing to give one, Newton was suggesting that gravity was not just unexplained but unexplainable—a magic trick we could never figure out. Leibniz wrote to Clarke: "That means of communication (says he) is invisible, intangible, not mechanical. He might as well have added, inexplicable, unintelligible, precarious, groundless and unexampled . . . 'Tis a chimerical thing, a scholastic occult quantity."

Newton openly admitted he didn't know how gravity worked: "I have not been able to discover the cause of those properties of gravity from phenomena, and I frame no hypotheses." He basically accepted Leibniz's jibe that gravity was "occult"—brought about by hidden causes—but didn't think it mattered. You may not know what causes gravity, but if you just accept its existence, almost every known fact about the universe falls into place, and that's good enough.

•

Following Newton's lead, modern physicists think of any theory as having two separate functions. First, the theory should provide a mathematical description: formulas that let you calculate how fast an apple falls, when the Moon will eclipse the Sun, or what have you. Second, the theory should provide an "interpretation" of the formulas: a compelling picture of what's going on with that apple or moon. For Leibniz and most other philosophers prior to Newton, the second was paramount. Their prime goal was to make the universe comprehensible. But with Newton, the first gained the upper hand. If you have to choose between description and explanation, physicists figure it's better to have description. By learning to live with your ignorance, you free yourself to make progress little by little. You can try to fill in the explanation later, and in the meantime you've got some handy formulas that convince your mom you're doing something useful with your life.

Modern physicists call interpretation a "philosophical" issue, implying that it requires a different frame of mind or a different academic discipline altogether. They spend their working hours doing calculations and can be at a loss for words when you ask them what's actually going on out there in the real world. If anything, they regard interpretation as suspect. An insistence on explanation had forced the atomists to come up with some idea, any idea, just for the sake of having a simple, comprehensible worldview. So perhaps it's better to focus on what we do know: observed facts. "Nothing is more requisite for a true philosopher, than to restrain the intemperate desire of searching into causes," wrote a prominent proponent of this view, the eighteenth-century Scottish philosopher David Hume.

Taken to extremes, this attitude is known as instrumentalism, which treats theories merely as mathematical tools, or instruments, for cataloging facts. "Shut up and calculate" is the instrumentalists' slogan.

This no-nonsense, just-the-facts-ma'am view of science cycles in and out of fashion. Widespread in the decades after Newton, it became popular again in the mid-nineteenth century, and again in the early- to mid-twentieth. Not coincidentally, these were periods of scientific revolution. When physicists introduce a controversial theory, they often, as Newton did, reassure their colleagues (and themselves) that it's really just a tool for doing calculations. If you can't fathom how the theory could be true, no worries—you don't need to believe it to use it. A spoonful of instrumentalism helps the radical idea go down.

Ultimately, though, instrumentalism is just a tactical retreat. In the end most people still crave a picture of what the universe is really like, what lies under the surface of our perceptions. After all, how can physics theories work so well if they don't capture some element of the truth? Young people, in particular, get frustrated when professors tell them not to worry their pretty little heads about what's really going on. A remarkable number of the most innovative scientists in history say they learned their chosen subject on their own because no one would teach it to them in class.

What's more, interpretation is not just an after-the-fact gloss on the equations, but the creative spark of science. After all, how do physicists come up with equations to begin with? Almost always, they have some specific physical picture in their head—in Newton's case, magical sympathies. Once physicists have developed equations on the basis of these mental images, they can kick away the interpretation and let the equations stand on their own, just as Newton distanced himself (at least in public) from magic. Any given set of equations has multiple interpretations, so physicists need not commit themselves to the one that got them there. They are free to come up with new interpretations, some of which will inspire new theories and new equations, and so the cycle continues. But they can never do without interpretation at all. There's no sharp boundary between philosophical issues and physical issues, merely a porous border with lots of trade across it.

•

Indeed, despite Newton's vow not to frame hypotheses for the workings of gravity, he did in fact frame a number of hypotheses—three broad categories of them, each of which gained adherents. First, maybe gravity involved a local, mechanical process after all. On the face of it, this option seemed like a nonstarter. Newton's law that the strength of gravity depends on the mass of an object surely dealt it a knockout blow. If the only way that particles impart a force is by colliding, their effects should depend on exterior surface area—how big a target an object presents—rather than on mass. Yet Newton still toyed with mechanical ideas, and one of his best friends, the Swiss mathematician Nicolas Fatio de Duillier, came up with an ingenious solution to the mass problem. If our planet were like a giant Wiffle Ball, riddled with tiny pores, particles from the outside could enter and collide with matter deep in the interior, and the force would indeed depend on the total amount of matter—that is, on mass. The theory failed to win converts less because of its failings than because of Fatio's: the man eventually managed to antagonize both Newton and Leibniz and fell in with a group of violent religious fanatics.

Second, maybe there was some way, besides collisions, for things to interact locally and produce a gravitational force. Newton and Clarke wrote of an "immaterial," "incorporeal," or "intangible" medium that would act as an intermediary, transmitting gravity from one object to another. These words have various connotations, such as God or spirits, but at their most basic they just mean something that does not consist of particles and does not respect the usual atomist rules about inhabiting space. Material particles are impenetrable; if one occupies a volume of space, nothing else can occupy the same volume. But an immaterial medium would not occupy a volume of space exclusively; it could intermingle with other things. Thus it could penetrate the interior of planets, explaining why the strength of gravity depends on mass rather than surface area. Leibniz, for his part, developed a theory of immaterial entities called "monads" that underlie our observed reality. Although he was never able to connect these monads to anything that can be seen directly, later philosophers such as Immanuel Kant picked up on the concept, and through them Leibniz's ideas inspired the concept of electric and magnetic fields.

Newton came tantalizingly close to identifying an immaterial medium capable of transmitting gravity: space itself. For him, space was a manifestation of God's omnipresence. He saw gravitation in the same way: the force leaps from one place to another because God already exists in both places. If gravity and space are associated with God's ubiquity, then gravity and space are associated with each other. Leibniz, too, implicitly associated gravitation with the nature of space. He thought his monads gave rise to our perceptions both of space and of action at a distance. To be sure, neither Newton nor Leibniz ever said that space caused gravity. Neither man thought space was capable of acting. It was Einstein who made that leap.

The above two interpretations say that gravity acts *as if* it were nonlocal. The third interpretation drops the "as if" and suggests that objects really do tug on one another across space. An early proponent of this option was Roger Cotes, an English mathematician who helped Newton revise the *Principia* for its second edition in 1713. Some historians think nonlocality may have been Newton's own favored idea, too. It's hard to tell. In one much-quoted letter, Newton appeared to call nonlocality "inconceivable . . . an absurdity," but when you read the quote in context, it might have been referring to atheism. In other writings, he merrily invoked nonlocal forces for lots of phenomena besides gravity, including light reflection and refraction, vapor diffusion, gas pressure, material cohesion, and heat. Newton never came out and declared that gravity is nonlocal, but that might have been because he didn't want to alienate mechanical purists even more than he already had.

•

Those who came of age in the decades after the *Principia* saw forces acting at a distance as perfectly reasonable. With notable exceptions, eighteenth-century scholars felt no need to conjure up some local explanation or make instrumentalist excuses. They extended the principles of gravity to the other examples of nonlocality that had cast a pall over physics. Benjamin Franklin, for example, put America on the scientific map by explaining electricity as a fluid of attracting and repelling particles, which were tacitly nonlocal. Others proposed such fluids for magnetism, chemical reactions, and much besides.

In fact, prevailing wisdom did a backflip. It was *locality* that started to seem unreasonable. Leave aside gravity, electricity, and magnetism: even the supposedly simple collision of two billiard balls made people smack their foreheads in dismay. Why do the balls rebound? Leading proponents of locality such as Democritus, Descartes, and Leibniz had wondered about that themselves. When the balls touch, are they still two balls, or have they become one? How does the effect propagate from the impact site to the other side of each ball? Do the balls really make an instant U-turn, which would entail an infinitely fast change of velocity?

Kant knew a thing or two about billiards. The eighteenth-century German überphilosopher played so well that his winnings helped pay his way through college. Kant was a key figure in sinking pre-Newtonian notions of locality. His overarching interest was to analyze how we know what we know, or think we know. Locality was an example of a familiar idea that, on closer inspection, is suspect. In daily life, we observe that we need to touch objects to get them to move. But the truth is that we never touch anything. Rather, we exert forces on them, and they on us. These forces account for the resistance we feel when we squeeze a ball or try to stick our arm through a solid wall. Familiar objects are actually mostly empty space. When we talk of "matter," we really mean a continuum of forces, as opposed to the constituent particles, which always remain inaccessible to us.

The original appeal of locality had been that a single mode of interaction—direct contact—could explain everything. When Newton added a second mode—nonlocal forces—he seemed at first to complicate matters, but Kant and others explained away contact and restored the earlier simplicity. If you watch in slow motion the collision of two billiard balls, you will see not a sharp rebound but a gradual reversal. As the balls approach, they exert a repulsive force on each other, which slows them down, stops them, and sends them back the way they came. They never actually come into contact. Whereas the old mechanical philosophers had sought to explain nonlocal forces in terms of local interactions, the new mechanical philosophers reduced local forces to nonlocal interactions.

Newtonian gravitation had come into the world struggling for acceptance, but it became the new orthodoxy. In 1872 the Austrian physicist-philosopher Ernst Mach described this turn of events. He argued that scientists explain phenomena by relating the unfamiliar to the familiar, the uncommon to the common. What is "common" might not really be any more intelligible than the uncommon, as we realize when a five-year-old asks how a common household appliance works and we fumble for an answer. But we're more comfortable accepting it. After all, we have to take *something* as the bottom level of reality, and it needs to be something that we can live with. Mach wrote: "Simplest facts, to which we reduce the more complicated ones, are always unintelligible in themselves, that is to say, they are not further resolvable . . . People usually reduce uncommon unintelligibilities to common ones."

But what we deem "common" can shift. Prior to Newton, it meant direct impact. After him, nonlocal forces came to seem common. "The Newtonian theory of gravitation, on its appearance, disturbed almost all investigators of nature because it was founded on an uncommon unintelligibility," Mach wrote. "People tried to reduce gravitation to pressure and impact. At the present day gravitation no longer disturbs anybody; it has become a *common* unintelligibility." How ironic: even before Mach penned these words, the pendulum had started to swing back, and physicists were again coming around to the idea that the universe had to be local after all.

Questioning Sir Isaac

The renaissance of locality had begun in 1786 with some dead frogs hanging from an iron railing. The Italian physician Luigi Galvani was doing experiments on how static electric shocks cause animal muscles to contract. One day he saw the frogs' legs twitch on their own, even when he wasn't deliberately zapping them, and it dawned on him that animal tissue didn't just respond to electricity, but could also generate it. Metal and amphibian formed what we now call a battery, and in 1800 another Italian, Alessandro Volta, built a practical one by substituting a wet piece of cardboard for the frog. Not only did batteries give experimenters an amazing new plaything, their very existence shocked those of a Newtonian frame of mind, who held that chemicals and electricity involved different types of nonlocal fluids and should not have been interconvertible.

The timing was fortuitous. Within philosophy, Kant's questioning of what our rational minds were able to grasp had stirred a rebellion against mechanistic thinking—a movement known as German Romanticism and, under that general rubric, a school of thought known as *Naturphilosophie*. *Naturphilosophie* represented one of the periodic revivals of magical thinking. Adherents were fascinated by Renaissance occult figures and Eastern mysticism. They considered nature's diverse forces, including both electricity and magnetism, different expressions of an organic unity. Practitioners of *Naturphilosophie* looked to experiment to discern this unity and make use of it for human needs. Their ranks included some of the greatest experimental scientists of the early nineteenth century.

Among them was the Danish experimenter Hans Christian Ørsted—not a doctor, as so many revolutionaries in science had been, but close: a pharmacist. He built his first battery shortly after hearing of Volta's invention and was soon hacking his own designs. At the time, experimental data backed up the Newtonian view that electricity and magnetism were unconnected. Static electricity causes no magnetic effects. But Ørsted figured that a flowing electric current might. And he was right. In 1820 Ørsted found that a wire connected to a battery can cause a nearby compass needle to pivot. Yet again a cozy consensus had been upset by ideas that most scientists had dismissed as unscientific.

Not only did Ørsted show that electricity and magnetism were related after all but that the nature of the relation was very un-Newtonian. The electric current did not push or pull on the compass needle, but spun it. That posed a serious challenge for nonlocal forces. Those forces are supposed to be like private hotlines between the two affected objects, acting as though the rest of the universe weren't even there. It stands to reason that the force should act along the straight line connecting the two, because it's blind to other objects or locations that might define an alternative direction. Ørsted's twitching compass violated this intuition. To spin the needle, the current had to be acting laterally rather than directly toward or away from the wire. As a further sign that a local rather than nonlocal process was involved, the spinning needle reminded people of the swirling motions that Democritus and Descartes had talked about as a mechanism for magnetism and gravitation.

While all this was going on, another branch of physics was undergoing its own upheaval: optics. Most everyone at the time accepted Newton's atomistic explanation of light as a stream of particles. But another doctor, Thomas Young in England, took his inspiration from the flow of water and other fluids. Like Aristotle, he thought light was an impulse conveyed through a medium that filled space. This explanation had been popular among medieval scholastics. In 1803 Young hit upon an experiment that would restore its popularity.

Imagine a sunny day, a window with a dark curtain, and a white wall opposite the window. If you cut a slit in the curtain, a little spot appears on the wall. If you cut a second slit in the curtain, the wall doesn't just show a second spot, as you might expect and as Newton's particle theory of light predicted. Instead, the wall is covered in a pattern like zebra stripes: alternating bright and dark bands called fringes. In fact, even though cutting a second slit lets more light through, the original spot typically gets *dimmer*. But it all makes sense if light is a wave in an invisible medium. The waves passing through the two slits overlap and can cancel out or mutually reinforce. In places where the peak of one wave coincides with the peak of the other, the wave becomes extra-strong, creating a bright band; where a peak coincides with a trough, the wave gets zeroed out, for a dark band. The effect is known as wave interference. Young's experiment is a physics classic. You can do it yourself (the trick is to make the holes as small as possible and use a laser pointer as your light source, rather than the Sun). In fact, nonlocality experiments like those I mentioned in chapter 1 are a fancy version of Young's experiment.

As compelling as it was, Young's idea languished for a decade and a half. The breakthrough was not a specific discovery, but the fall of Napoleon. Under the emperor, Newtonian scholars in France had shut out rival theories. Even in England, people were apt to misinterpret Young's work. Only when the Newtonians' political and intellectual authority weakened did latent interest in the wave nature of light come

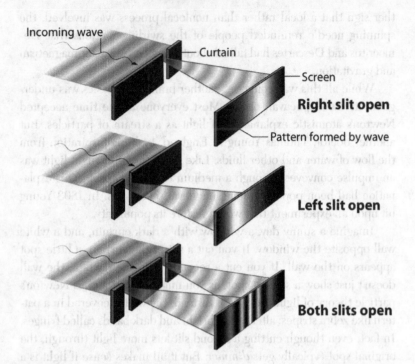

Incoming wave

Curtain

Screen

Right slit open

Pattern formed by wave

Left slit open

Both slits open

2.1. Two-slit experiment. If you open a slit in a dark curtain, light waves will illuminate a spot on a screen. If you open two slits, the waves will overlap and create a striped pattern, called "interference fringes." (Illustration by Jen Christensen)

into the open. General opinion swung in its favor by the 1820s—just as Ørsted was revolutionizing the study of electricity and magnetism.

These two challenges to Newtonian theory met in the person of Michael Faraday. Faraday is one of the most fascinating figures in the history of science, an example of how inquiry benefits from diversity. Born into a poor London family, Faraday barely went to school. He became a bookbinder's apprentice and got interested in science while reading a volume of the *Encyclopedia Britannica* that a customer had left in the shop. He borrowed a shilling from his brother to attend a science lecture and built his own battery on the mantelpiece at the back of the shop. Before long he talked himself into a job with the most famous chemist in Britain, Humphry Davy, who had spent time in Germany with the Romantics there and shared their vision of unity in nature.

Faraday became a leading figure in physics just as physics was becoming *physics* as opposed to a branch of philosophy. The word "physicist" was coined in 1840. When you ask scientists today, they differentiate between physics and philosophy based on the importance of experiments. But historically the schism was a rebranding strategy, part of the general standardization and professionalization of academic disciplines in the nineteenth century.

Faraday never learned math, and it's a good thing for the rest of us that he didn't. The mathematical elegance of Newtonian theory meant nothing to him, so he felt completely free to explore radical concepts. For him, the most straightforward interpretation of Ørsted's discovery was that nature is local after all. Equally, though, Faraday recognized that scientists couldn't turn the clock back to atomist theories in which objects influence one another only by colliding. There had to be some other way for things to interact locally.

He thought the light theorists were onto something with their notion of influences sweeping through an all-pervading medium. Although electromagnetism seemed like an utterly different phenomenon from light, it, too, hinted at a medium. If you sprinkle iron filings onto a magnet, they arrange themselves into graceful arcs called lines of force, which bear an uncanny resemblance to the strain pattern that develops in any elastic material when you stretch it. To Faraday, the filings were like dark soot collecting on the body of an invisible man: they gave away the medium's presence.

But what kind of medium could this be? Faraday originally pictured it as an ordinary substance made of little particles, each individually obeying Newton's laws of motion. But it gradually dawned on him that the electromagnetic medium could be no ordinary substance. For one thing, whereas only one ordinary object can occupy a given space at a time, this medium coexisted with other things. The arcs formed by iron filings don't stop at the poles of a magnet, but continue through the body of the magnet and close back on themselves in a loop; the lines of force thread through matter, existing independently of it. So Faraday and others imagined the medium as a new category of stuff—an immaterial medium or continuum of force like the one that Newton, Leibniz, Kant, and others once speculated about. In 1845 Faraday introduced the term by which we know this medium today: the "field."

The field surrounds us and infuses us; we are swimming in it and it is always tugging on us. We never see it directly, but it makes its presence felt by communicating forces from one place to another. The field is local in two senses. First, an electromagnet does not magically reach across space to tug on a metal paper clip. The paper clip is affected only by the condition of the field at its location, just as a water bug can be placidly floating on the surface of a pond, oblivious to children splashing in the water on the other shore. Second, the electromagnet takes time to exert its effect. When you first turn it on, the paper clip does not instantly feel a force. The effect must sweep through the field until it reaches the paper clip and causes it to snap toward the magnet, just as splashing in the water sends ripples across the pond, eventually swamping the poor bug. The same logic applies to electric forces. If you rub a latex balloon on your sleeve and hold it next to your head, it does not instantly tousle your hair. Rather, it disturbs the electric field and the effects spread across the gap between balloon and hair, eventually altering the condition of the field at your scalp.

•

Faraday's field concept failed to catch on at first. Skeptics insisted on seeing formulas, and Faraday, the mathematical illiterate, had none to offer. But his ideas electrified a younger generation of mathematical whizzes, notably the Scottish physicist James Clerk Maxwell, who turned Faraday's intuitions into equations. To capture fields mathematically, Maxwell used a system familiar to anyone who has watched a weather report. A weather map displays lots of numbers and little arrows that tell you the temperature, wind speed, wind direction, and so on at various sites. Likewise, Maxwell represented the electric and magnetic fields by little arrows that indicate the strength and direction of the field at points in space. Grids of numbers tell you how the field pushes on electrically charged objects or magnetic compass needles. Maxwell's famous equations predict how these quantities change over time.

Today you can buy T-shirts with the equations silk-screened onto them. They are the epitome of the elegant theory to which all physicists aspire. In addition to the magnetic and electric fields, the universe is laced by dozens of other intermingling fields, corresponding to the sundry forces of nature. For all Maxwell's success, though, the meaning of his equations was hazy. Did they really respect the principle of locality? They seemed to, but appearances can deceive. For one thing, although Maxwell had designed his equations to depict forces acting locally, he admitted they could equally well portray forces acting nonlocally. On that interpretation, space would not be filled with a tangible medium; it would be mostly empty, with some objects scattered here and there, pulling and pushing on one another from afar. The numbers assigned to spatial points would describe a hypothetical: If you placed an object at such-and-such a location, how would all the other objects in the universe act on it? Maxwell's theory therefore sparked the same kind of interpretational debate as Newton's law of gravity had done two centuries earlier.

Three features of fields affirmed they are real. First, fields have a life of their own. They are not merely middlemen conveying impulses from one object to another. They can rouse themselves into action, independent of matter; a space completely devoid of particles can still be humming with wave activity. This phenomenon is alien to the nonlocal picture. Second, electrical and magnetic disturbances take time to exert their effects. A time lag seems odd if forces are leaping directly from one object to another, but is perfectly natural if an impulse must make its way through a medium. In fact, the speed at which these effects propagate equals the speed of light. Evidently, light is an electromagnetic wave. Finally, fields contain energy, the very essence of real things (and a fairly new concept in physics at the time). Their ability to store energy ensures that none goes missing in the time it takes for a disturbance to propagate through space.

These three criteria—waves, time delay, energy—persuaded most of Maxwell's contemporaries that fields provide a local explanation for electrical and magnetic forces. Conventional wisdom did yet another

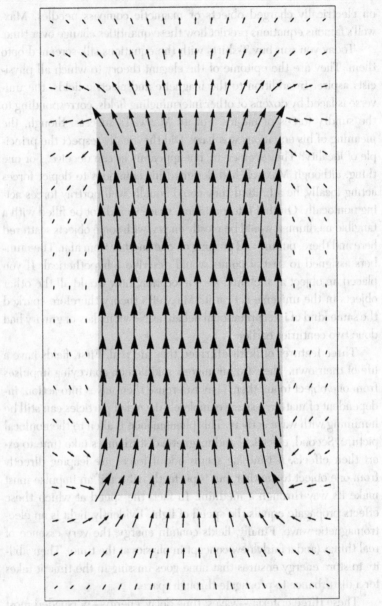

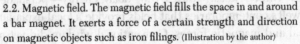
2.2. Magnetic field. The magnetic field fills the space in and around a bar magnet. It exerts a force of a certain strength and direction on magnetic objects such as iron filings. (Illustration by the author)

backflip: nonlocality went from orthodoxy to "a very old but most pernicious heresy" and "unthinkable." In the sweep of history, these remarks
have a familiar ring. Here again was a generation of physicists making
confident declarations that contradicted the confident declarations of
earlier generations. In fact, the bravado hid some unease.

Renewed Troubles for Locality

Fin-de-siècle physicists fretted that they had two separate theories:
electromagnetism and mechanics. A big jagged crack ran down their
view of the world, which not only spoiled their dream of simplicity, but
also left them at a loss to solve various practical problems. To track
baseballs and planets, they applied Newton's laws. To build generators
and electromagnets, they applied Maxwell's equations. But what were
they to do in situations that mix up motion with electromagnetism?
How does a moving object affect electric and magnetic fields, and
vice versa?

The two theories seemed downright incompatible. One of the central features of Newton's laws, gravity, had no place in Maxwell's theory.
Whereas electrical and magnetic forces can either push or pull, gravity
always pulls. Also, the gravitational field satisfied none of the criteria
that had demonstrated the reality of electric and magnetic fields. For
instance, observers saw no sign that gravity took time to propagate.
According to an influential (though, in retrospect, wrong) estimate,
gravity zipped across space instantaneously. At the beginning of the
century, gravity had been a model for the other forces; by the end, it
was a disconcerting outlier.

An even more basic problem was that Maxwell's equations single
out a certain speed as special—the speed of light—yet Newton's laws
say there is no such thing as "the" speed of anything. In those laws,
speed is always relative. Relative to the person who throws it, a baseball might be flying at 20 mph; relative to someone watching from a
moving train, 100 mph; relative to an astronaut on the space station,
17,000 mph. If, instead of throwing a ball, the person shines a flashlight,
how fast are the light waves moving relative to those onlookers? Years

later, one physicist would recall reading about electromagnetism at age sixteen and wondering: If you moved at the speed of light, would the waves look like they're standing still? Some theorists thought yes, others no. Experiments were equally contradictory.

One theorist who puzzled over the incompatibility of mechanics and electromagnetism was the Dutchman Hendrik Lorentz. His daughter Geertruida, who grew up to be a respected physicist in her own right, reminisced that she and her siblings teasingly called their father a polar bear, on account of how he rhythmically paced back and forth in his basement study like a bear in a cage. During these ursine perambulations, he came up with a way to reconcile mechanics and electromagnetism. He argued that electromagnetism is actually the deeper of the two theories. It could account for Newton's laws of motion and perhaps even of gravity. In Lorentz's approach, there *was* such a thing as "the" speed of objects, determined by the electromagnetic medium. If you're moving at waves' native speed, they'll appear motionless.

Lorentz had a ready reply to experimental results that suggested otherwise: physicists were being fooled by the experimental design. To measure speed, they needed a yardstick, and they needed to trust that the yardstick would provide a reliable standard of length. This trust was misplaced. Lorentz reasoned that when a yardstick is in motion, the electromagnetic field will resist that motion, squeezing it lengthwise like a falling raindrop flattened by the air drag. This effect will throw off the measurement and mislead experimenters into thinking that the speed of light relative to the apparatus never varies. In short, although you have some true, absolute velocity, electromagnetism foils your attempts to measure it.

On a purely practical level, Lorentz's theory was a huge success, and in 1902 earned him a newly minted honor: the Nobel prize. But it did seem rather sneaky of nature to play such tricks on experimenters. And the theory created some new problems of its own. Physicists and philosophers had been vacillating for thousands of years between discrete particles and continuous media, but Lorentz combined *both* into his theory, with awkward consequences. For instance, the electric field should convey influences not just from each charged particle to every other, but also from each particle to itself. This self-referential loop would create paradoxes. A particle would begin to accelerate *before* you applied a force to it, as though it were psychic. The ability of a particle to see slightly into its own future could be used to send messages from one place to another infinitely fast.

As if that weren't enough, the theory predicted that particles would explode under the pressure of their own pent-up electric field. To explain why the particles of the universe aren't popping off like fire-crackers, physicists reckoned that they must be true geometric points of zero size. Nothing so insubstantial would be capable of blowing up. But as Zeno had pointed out two thousand years earlier, a point is a paradoxical thing. Any time the number zero appears in physics, the number infinity can't be far behind. If the electric field is focused to an infinitesimal point, it becomes infinitely strong. For analogous reasons, if the wavelength of light can be any number all the way down to zero, a box filled with light waves would have an infinite capacity to store energy. Such a box would suck in energy like a black hole, not because of its gravitational force, but because of its unlimited storage capacity—like people on the TV show *Hoarders* who are such pack rats that things disappear into their house, never to be seen again.

In short, whenever physicists attempted to describe particles inter-acting locally, whether by colliding or by sending out ripples in fields, they kept encountering the word "infinity." Some physicists came to question not just Lorentz's theory but the concept of fields and the principle of locality. The trouble that nineteenth-century physicists were having with the disunity of their subject closely mirrors the situa-tion we face today, in which theorists struggle to reconcile gravity with the other forces of nature. Soon, that young man who had posed the question about light waves at age sixteen would come of age and sort this mess out.

3

Einstein's Locality

As a college student, Einstein blew off a lot of his classes. He didn't think much of how physics was taught. His professors left out all the fun stuff, not least the ferment that Maxwell's theory of electromagnetism had stirred. Einstein spent much of his time sitting in Café Metropole in Zurich and delving into the great works of philosophy: Hume, Kant, Mach. If not for his friends' class notes, he might never have graduated. For their part, his professors found him a bit too full of himself and gave him negative job references. Laboratory directors across Europe would later have to live down the fact they had rejected job applications from Albert Einstein, of all people.

In the earliest years of his career, Einstein didn't attach much importance to locality. He was a Newtonian. For his first scientific papers, he assumed that particles acted on one another at a distance. If Newton's laws clashed with Maxwell's equations, so much the worse for Maxwell. In particular, if Newton's laws hold that all velocities are relative, then the velocity of light had to be relative, whatever Maxwell's equations might suggest. So Einstein tweaked those equations to make the speed of light relative to its source, creating a new version of electromagnetic theory that was nonlocal. And that's when he had a change of heart. The revised theory did so much violence to Maxwell's original version that experiments appeared to rule it out. What is more, it predicted that some people would see electromagnetism obey Maxwell's equations in their native form, while others would see a distorted version—a prospect that offended Einstein's egalitarian instincts.

In a quintessential eureka moment, Einstein realized that velocities can be relative *and* light can set an absolute velocity standard. There's no contradiction, as everyone had thought. You just have to be careful about what a relative velocity is. The usual rule embodied in Newton's laws is that you calculate relative velocity by addition or subtraction: a 20-mph baseball thrown at an oncoming 80-mph train is moving at 100 mph relative to a passenger. Yet this rule contains a tacit and unwarranted assumption of instantaneous communication or, equivalently, of nonlocality.

Einstein realized this when he thought about what is actually involved in comparing velocities. He used—indeed, he pioneered—one of modern physicists' favorite styles of reasoning, called "operational" reasoning, in which you ask *how* you know what you know. Often you discover you had no justification for your beliefs and that those beliefs are in fact wrong. By the way, this technique can make progress on all sorts of disputes. To elevate a political debate, ask *how* something happens. For instance, if someone supports or opposes single-payer health care, ask how health insurance actually works. People who are very sure of their opinions will be forced to confront their own ignorance or at least admit that the issue isn't so clear-cut.

In the case of relative velocities, Einstein pointed out that, for the ball-thrower and train passenger to measure the ball's velocity, each needs a stopwatch. And they can't take for granted that their watches keep the same time. That's something they need to establish by comparing clock readings, which requires them to exchange some kind of signal. If the signal zaps between them instantly, they can confirm that an hour for one person is the same as an hour for the other. But if the signal transmission takes time, they can't be so sure, because they will shift position while the signal is en route, creating a lag. Nor do they really know that a mile for one is a mile for the other. A length measurement is implicitly conducted at one instant in time, which assumes a signal can travel instantaneously from one end of an object to the other. If observers are restricted to signals moving at a limited speed, the measurement could be thrown off by their or the object's motion.

Einstein found an alternative to the Newtonian velocity-addition rule that takes the signal transmission time into account while still ensuring that the ball-thrower's perspective and the train passenger's perspective are perfectly equivalent. By this rule, the combined velocity is *less* than the straight sum. For the passenger, the ball is moving ever so slightly slower than 100 mph. The faster the ball is thrown, the more its relative velocity deviates from the Newtonian expectation. If, as nineteenth-century theorists had wondered, the person on the ground shines a flashlight rather than throwing a ball, the light waves will be moving at 670 million mph relative to him and 670 million mph relative to the passenger. The passenger's own motion ceases to matter. Thus light moves at a velocity that all observers agree upon, even though observers' own velocities are always relative.

Einstein's rethinking of relative velocity accounted for all the experiments that puzzled his contemporaries. The discrepancies evaporated; there was no need to suspect that nature was maliciously foiling experimenters. These successes disabused him of his earlier willingness to countenance nonlocality.

•

By confirming that all velocities are relative, Einstein relieved the main tension between the laws of motion and of electromagnetism. Even those college professors who had found Einstein maddening as a student were impressed, and one of them pointed out some repercussions the impudent young genius had missed. When you start mucking with a basic concept like speed, you do a lot more than solve one puzzle. Speed is defined within space and time, so Einstein's reinterpretation of it transformed what physicists mean by those concepts. Because people moving at different speeds can't keep their watches in sync, time intervals depend on their speed; so do spatial distances, for related reasons. But the combination of time interval and spatial distance—the spatiotemporal "distance"—does not depend on speed; it's an objective fact that everyone can agree on. This is how Einstein's theory of relativity achieves its renowned unification of space and time into a single concept, spacetime. To today's physicists, this union is the real meaning of the theory, and the business with trains and signals was just one means to discover it.

We still perceive spacetime as space and time, but no person has a monopoly on the division of spacetime into "space" and "time." What is purely spatial for one observer is a combination of spatial and temporal for another. To a train passenger, the newspaper on his or her lap is "here" (a purely spatial designation), but to someone watching from the ground, the newspaper is a moving target (a mix of spatial and temporal). Those two people also have different concepts of "now" and will disagree on what events occur at the same time. The phrase "at the same time" is a four-letter word in relativity theory—objectively there is no such thing.

One puzzle piece still didn't fit in: the force of gravity. Relativity theory, in its original version, applies only in the special case of zero gravity. In 1915 Einstein rounded out the picture with his general theory of relativity, according to which gravity is produced by a field analogous to the electromagnetic field. The arc of a baseball in flight isn't caused by a force that Earth exerts on the ball at a distance, as Newton's theory held. Instead, the ball is responding to the gravitational field immediately around it. When Earth's mass shifts around—for instance, when geologic activity or ocean currents redistribute material—the gravitational field changes slightly. This disturbance ripples outward through the field at the speed of light and, as it passes through the baseball field, it reshapes the gravitational field there, so that the next time you throw a ball, it might fall ever so slightly faster or slower.

The gravitational field is not just any field, though. It plays a special role in nature. All other fields are selective: the electromagnetic field, for example, acts only on electrically charged objects, and the more strongly charged an object is, the faster it will accelerate. In contrast, the gravitational field acts equally on all objects. Everything accelerates downward at the same rate. The field thereby marks out the path that all objects take in the absence of other forces. But that's the very function space has. So the gravitational field, according to Einstein, is not located *in* space, but is a property *of* space. If the fabric of spacetime is like a rug and a moving object like a marble rolling across the rug, then Earth's gravitational field is a bump that diverts the marble in a new direction.

Just as a plot twist in a novel doesn't change the story up to that point, but can still cause you to reassess earlier events—a character you thought was evil might actually be the good guy—so did general relativity prompt physicists to reappraise Newton's theory of gravity. His theory isn't strictly wrong, but incomplete. It roughly captures the effects that gravity has, but fails to account for how the force propagates. Relativity fleshes it out. It vindicates Newton's and Leibniz's foggy intuitions that gravity has something to do with the nature of space.

Locality Keeps Physics Sane

Einstein developed relativity theory by thinking about what locality means for our measurements of time and length. In turn, his theory contains provisions that enforce locality. Above all, relativity theory implies that nothing can travel faster than light. Technically, the theory doesn't forbid faster-than-light motion per se. It says merely that light moves at the same speed for all observers. Under most circumstances, though, this requirement translates into a universal speed limit. If you could catch up to light, then, as the teenage Einstein mused, light would appear to be standing still—it would no longer be moving at the same speed relative to you as to everyone else. No matter how fast you go, how earnestly you try, you can't catch up. It is as futile as seeking the end of a rainbow.

In practice, if you try to accelerate an object to light speed, a kind of governor rides the brakes, so that you have to work harder for each additional increment of speed. That is why modern particle accelerators have to be so ginormous. The hairline difference between 99.9999 percent of the speed of light (the speed of particles in the old Tevatron accelerator at Fermilab) and 99.999999 percent (the speed in the Large Hadron Collider) is a factor of 10 in energy. Reaching light speed would take an infinite amount of energy.

•

The universal speed limit rules out the infinitely fast nonlocal forces that Newton postulated. What is more, like an enlightened parent who doesn't just make household rules but spells out their rationale, relativity theory doesn't just ban faster-than-light travel but clarifies exactly why it would be so troublesome.

First and foremost, breaking the speed limit would muck up sequences of cause and effect. Different people would disagree not only on what "now" is, but on what is "before" and "after." To see why, go back to Einstein's operational mode of reasoning and ask yourself: How do I know the order that events occur in? You need to observe those events using light or some other probe that takes time to pass through space. If events occur in such rapid succession that light cannot pass between them in time, observations of those events could conflict, and people will disagree not just on the pace of events, but on the essence of what happened.

For instance, go back to the train scenario and, this time, imagine you throw a ball faster than light at a train that is moving away from you. The ball overtakes the train, punches a hole through the rear car, flies the length of the train, and exits the front. Or at least that's how it looks to you. A passenger on the train might see something different. It takes time for light from the ball's violent entry and exit to reach the passenger's eyes, and in the interim, the train continues moving forward, so that light coming from the rear of the train has to cover a longer distance, and light from the front has a shorter trip. Consequently, the passenger could see the ball punch through the front of the train *before* it punches through the rear. In fact, the whole sequence of events would be reversed: the ball flies backward, out the rear of the train, and into your waiting hands. Even if the passenger recognizes that appearances can deceive and makes allowances for the light transmission time, she thinks the sequence occurs in reverse order. And because the passenger's perspective is on a completely equal footing with your own, both of you are right. When things travel faster than light, the order of events is objectively ambiguous.

Such a reversal of causal sequences is not just mind-blowing but theory-busting. It amounts to moving backward in time. By relaying signals via observers who are moving faster than light or communicating nonlocally, you could send a message into your own past. Einstein realized this as early as 1907. "Using hyperlight velocities we could telegraph into the past," he remarked at a conference. He seemed sanguine about this prospect, but science-fiction writers knew better. They'd already based plots on the Pandora's box that time travel opens. In perhaps the earliest example, from 1881, the American writer Edward Page Mitchell told the story of a time traveler who saves the besieged sixteenth-century Dutch city of Leiden using his historical knowledge of what happened, thereby creating a causal loop: the traveler was responsible for the very event he remembers. Later stories toyed with going back in time and preventing your own birth by killing your grandfather or another ancestor. In a case of life imitating art, physicists and philosophers came to see time travel as an outright impossibility. If the laws of physics do nothing else, they should at least forestall logical contradictions. The universal speed limit does that.

In addition to maintaining the direction of causality, the speed limit ensures that the very concept of a law of physics makes sense. If objects or forces could move infinitely fast, the world would descend into anarchy. A dramatic example was discovered by Paul Painlevé, a man who, as minister of war and prime minister of France during an especially dark phase of the First World War, was rather well acquainted with anarchy. In the halcyon days of the mid-1890s, Painlevé had been a humble mathematician. One of his projects was to apply Newton's law of gravity to dense clusters of stars. Painlevé realized that the stars, buzzing around one another like bees in a hive, could whip themselves into such a frenzy that the laws of physics would be unable to say what happens next—a showstopper known as a singularity.

A singularity is any place or event where quantities become infinite and the mechanisms of nature blow a fuse. The center of a black hole is another example, albeit for different reasons. In Painlevé's case, one of the stars could fly off into deep space at infinite velocity. That's bad enough, but what's even worse is that the same could happen in reverse. At any moment a new star might swoop in from infinitely far away—a space invader, as one philosopher later put it. If so, then the laws of physics couldn't predict for sure what would happen to the cluster or, indeed, to anything. A space invader could zip across the universe, steal the socks from your laundry, and return home before you knew it. This is much worse than other instances of randomness in physics, because you can't even assign probabilities to the possible outcomes.

The seeming absurdity is not limited to Newton's laws. Certain types of fields, too, allow pulses to ripple through them at infinite speed. Whenever speed is infinite, spatial distance loses all meaning, and nature becomes indeterministic, governed by caprice rather than law. Everything that physicists have been trying to build since ancient Greece would lie in ruins. By imposing a speed limit, relativity theory restores law and order, not to mention safeguarding your laundry basket. As Steve Giddings, the ice-climbing theorist, puts it, "Anything that's nonlocal has the potential to be garbage." Relativity vindicates the age-old intuition that nonlocality would render the universe incomprehensible.

•

Let's take stock of where things stood on the eve of the quantum revolution. In ancient times, philosophers convinced themselves that objects interact only by smacking into one another. In Newton's day, they persuaded themselves that contact action actually made no sense and that things had to act on one another at a distance. Then, beginning with Michael Faraday and culminating with Einstein, they had another change of heart and decided that things had to interact locally after all. Even Niels Bohr, who disagreed with Einstein about much else, called action at a distance "irrational" and "completely incomprehensible."

The atomists identified two aspects of locality. The principle of local action holds that influences do not jump from one place to another, but pass through all the intervening points. And the principle of separability says each distinct object has an independent reality. The world has structure; things do not melt together into some undifferentiated goop. The electromagnetic, gravitational, and other fields embody both of these concepts, albeit in ways the ancient atomists never imagined. Objects can interact not just by contact action, but also by continuous action—ripples in fields. Each and every point of a field is a distinct object in its own right, whose existence is an objective fact on which all observers will agree.

Whereas the Greeks and later mechanical philosophers held locality to be self-evident, Einstein justified it retroactively. As essential as he deemed it, he was careful to admit that it was a supposition and had to be judged by the empirical success of the framework of which it is a part. And the framework was undoubtedly successful. The satisfying way field theories clicked together vindicated locality in both of its aspects.

So, physicists really thought they had it this time. At the very moment of its triumph, however, the principle of locality—and, with it, the entire classical conception of space—came under renewed assault from the emerging theory of quantum mechanics. An idea that philosophers and physicists had fought for two and a half millennia to keep out—that influences could operate outside the shackles of space—forced its way in after all. Because Einstein had so deeply embedded locality in science, its new woes shook the foundations of science more violently than ever, and physicists are still picking up the pieces.

The Peculiar Genesis of Quantum Mechanics

"Quantum mechanics was conceived in sin," says the historian and philosopher Arthur Fine at the University of Washington. Whereas the physicists on a TV show such as *The Big Bang Theory* are geeky figures of fun, the fathers of quantum mechanics were epic, often tortured souls. Some of them struggled with depression; one committed suicide. One lived with both wife and mistress under the same roof. One joined the Nazi stormtroopers. They were not above misrepresenting their rivals' arguments to make a point. They admitted they didn't know what they were doing half the time. With beginnings like this, is it any wonder that the disputes rumble on?

Einstein was the central figure in this crazy drama. Textbooks typically flatten his contribution to quantum mechanics to a single discovery, known as the photoelectric effect, for which he received his Nobel in 1921. But he can fairly be called the theory's father and, for a decade, practically the only one who believed in it. His goal, at first, was to understand light. From Democritus to Aristotle to Newton to Thomas Young, theorists had ping-ponged between thinking of light as wave or as particle. In a series of papers beginning in 1905, Einstein settled the issue: it's *both*. That makes as much sense as a vegan butcher. How can light be both spread out in a smooth undulation and packed into local-ized wads of energy?

Leaving aside the apparent contradiction in terms, the dual na-ture of light posed a specific problem: it conflicted with the principle of locality. If light were *either* a particle or a wave, there'd be no trou-ble. Particles bounce around and interact by direct contact or, per-haps, short-range forces; waves ripple through a medium or field in a continuous motion. Atomists plumped for particles, field theorists were wild about waves, but all agreed that light was local. But when light acts as *both* wave and particle, nonlocality seems unavoidable. The reason is that it takes a high degree of coordination across space for these two kinds of behavior to dovetail. Einstein and other theorists didn't realize this nonlocality straight away. They took it for granted that nature was local; indeed, they considered the restoration of local-ity as the greatest lesson of nineteenth-century physics, enshrined in the theory of relativity. But nonlocality crept into their awareness as they tried and failed to fit the dual behavior of light into one of the old frameworks.

For instance, suppose light is ultimately a wave, but gives the impres-sion of being particulate because atoms absorb wave energy in discrete bites. Most of Einstein's contemporaries adopted this picture. But Einstein saw very early on that it ran afoul of what the physicist John Cramer of the University of Washington has called the "bubble para-dox." A wave would spread out from its source like an inflating bubble. Upon reaching an atom, the bubble would pop—the wave would col-lapse and concentrate all its energy in that one place, like an ocean wave breaking in a narrow cove. By that point, the bubble could be huge; the collapse would happen abruptly over a wide region. How would distant parts of the bubble know that they should cease propagating outward? Some mysterious nonlocal effect would have to be operating.

Alternatively, suppose light is ultimately a particle. If light some-times looks like a wave, that's just because the particles are undulating in unison, like spectators doing the wave in a stadium. This was Einstein's initial instinct. But he quickly realized this would contradict observa-tions. Light particles acting independently could explain the short-wavelength end of the light spectrum, but not the long-wavelength

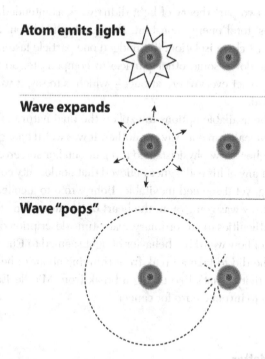

Atom emits light

Wave expands

Wave "pops"

3.1. Einstein's bubble paradox. Einstein devised this paradox in 1909 to argue that atoms emit light as discrete particles rather than continuous waves. A continuous wave would expand outward from an atom in a growing sphere like an inflating soap bubble. When the wave hit another atom, it would pop—the energy spread around the circumference would be focused in that one place. That would be a nonlocal process, which Einstein and his contemporaries thought implausible. It makes more sense to say that the first atom emits a particle in the direction of the second atom. Later Einstein extended the paradox from light waves to quantum wavefunctions. (Illustration by Jen Christensen)

side. At long wavelengths, the particles can't be independent; some external influence must be causing them to wiggle in unison.

As a third possibility, Einstein conjectured that light has two separate components, one a particle and one a wave, each of which behaves locally. Particles carry the radiative energy, while a ghostly "guiding field," with no energy of its own, sweeps those particles along like an ocean wave carrying a surfer. Bohr toyed with a version of this idea, too. Yet the two-part theory of light didn't work as intended. To keep the particles' total energy constant, the guiding field couldn't shepherd each particle in isolation; if it nudged one particle faster, it would have to slow down some other particle to compensate. So the wave would have to act everywhere at once—which is to say, it would have to be nonlocal.

All of the available options to explain the dual nature of light demanded nonlocality in one way or another. It was weird enough to have a vegan butcher. Now physicists had a vegan butcher sorcerer. Neither Einstein nor any of his colleagues believed that nonlocality could really be operating, yet it seemed inevitable. Bohr wrote to a colleague that something fishy was going on in the heart of the atom, something that "presents difficulties to our ordinary space-time description of nature." For a sense of how weird the behavior of light seemed to Einstein, consider what he did to take a break from thinking about it: he invented general relativity. That's like taking a break from Middle East peace negotiations to invent a cure for cancer.

Waves of Matter

In the early 1920s Einstein and younger colleagues such as Erwin Schrödinger took a crucial leap. They proposed that *all* forms of energy and matter, not just light, can behave as both particle and wave. The confusion over light now infected matter as well. Whether you give primacy to waves or to particles, you encounter nonlocality.

Schrödinger was a waves-firster. "Particles are nothing more than a kind of 'foamy crest' on wave radiation that constitutes the underlying basis of everything," he suggested. His thinking was that if widely separated particles are riding the same wave, they'll naturally stay synchronized—no nonlocal influences needed. Running with this idea, Schrödinger came up with the equation, now known simply as the Schrödinger equation, that is taught to physics students as the very definition of quantum mechanics. It lets you do everything from tracking a particle's motion to calculating the colors of light that an atom will emit or absorb. But, to his dismay, Schrödinger realized that his equation does not describe a wave but a "wavefunction," a curious mathematical abstraction that encodes the qualities of particles and systems of particles. The wavefunction is nonlocal. An entire swarm of particles has *one* wavefunction, binding together all their fates, no matter how dispersed they may be. Even a single particle's wavefunction spans the entire universe.

Schrödinger's rival, the German physicist Werner Heisenberg, a protégé of Bohr, leaned toward a particles-first theory and came up with his own set of equations. They proved to be mathematically equivalent to Schrödinger's; the two men had arrived at the same theory by two routes. But Heisenberg's version did little to clarify what was really going on. Heisenberg admitted he didn't know how his particle equations accounted for wave effects. Later, physicists realized that particles in Heisenberg's approach are wavy because they can react not just to what goes on immediately around them, but also to distant regions of space.

In short, both Schrödinger and Heisenberg had not demystified nonlocality, but heightened it. Indeed, the theory had a scrappy feel to it. Whereas the theory of relativity flowed organically from a single compelling principle (of symmetry) and quickly became accepted, quantum mechanics was cobbled together from various disconnected insights, and physicists had to reverse-engineer the essential principles it embodied. The situation was very similar to the puzzlement over Newton's theory of gravity two centuries earlier. To make sense of what they had wrought, physicists had to go beyond the equations and reach into their intuition for how the world should work. That's when the gloves really came off.

Einstein and Schrödinger took a position analogous to Newton's critics such as Leibniz: because the theory predicted nonlocality, it had to be provisional. The theory wasn't wrong, but incomplete. There must be some deeper theory that explains away the nonlocality. Bohr and Heisenberg argued that, no, the theory was not provisional. It was the final word in physics. To be sure, Einstein and Schrödinger were not in complete agreement, and Bohr and Heisenberg didn't always see eye to eye, either. But it's still fair to talk about two sides to the debate, if only because that's how the scientists themselves perceived it.

Bohr's and Heisenberg's views evolved into the so-called Copenhagen Interpretation. One of its central tenets is that nature is inherently random. The rationale for this was partly empirical. Quantum processes look random; for instance, when an atom spits out a photon, no known law determines the timing and direction of the emission. But the Copenhagenists went further than the data strictly demanded and argued that no such law is possible, period. The wavefunction specifies the odds of finding a particle in a given position or moving with a given velocity. Until someone goes to look for that particle, it exists in a state of limbo, having no definite location or momentum, but pregnant with possibility. In measuring its position, the experimenter causes the wavefunction to collapse down to a narrow spike located randomly within the range of possibilities, and the particle to pop up in the designated place. Collapse is abrupt and inexplicable, lying beyond the scope of Schrödinger's and Heisenberg's equations. As one philosopher has written, "This collapse is, quite literally, a miracle."

In their defense, seeking causes for the random vagaries of life is a recipe for frustration. Some things just happen. Misfortune befalls the virtuous, while the wicked prosper. For Copenhagenists, indeterminism was a lesson of modernity, an antidote to a misplaced Enlightenment trust in reason, which German intellectuals in the 1920s widely held responsible for their country's defeat in the First World War. A number of historians have traced this cultural mood to the magical and Romantic belief that nature is beyond rational understanding.

Einstein and Schrödinger loathed this reading of quantum mechanics. God does not throw dice, Einstein famously said. This highly quotable remark makes it sound as though Einstein had a religious aversion to indeterminism. The truth, as always, is rather more complicated and interesting. Einstein never objected to randomness per se; he had spent much of his career studying random processes. His concerns were more pragmatic. Other random phenomena arose from finer-scale mechanical motions. Why should quantum randomness be any different? Why should physicists give up the search for a deeper level of nature? As I mentioned in an earlier chapter, Einstein was struck by the fact that the universe is comprehensible in so many ways, and he thought it strange to suppose that particles would prove to be an exception. Either the universe should be intelligible or it should be inscrutable, but not half and half.

Furthermore, Einstein saw that indeterminism would entail non-locality. The reason is that the supposedly random events in quantum mechanics are coordinated. Not only do we observe them to be coordinated, but they *must* be coordinated or else energy or momentum will be lost or gained. For instance, those magic coins I described in chapter 1 land randomly on heads or tails, but do so in lockstep. How do they do that? If the outcome of the tosses is decided on the fly, the coins must be communicating nonlocally to ensure they match. Conversely, if they're not communicating, the outcome must have been preordained, and quantum theory, by failing to specify this outcome, must be incomplete. Over the years Einstein refined this dilemma between indeterminism with nonlocality or determinism with locality.

The 1927 Debate

The first time Einstein presented his dilemma was in October 1927 at a conference that ranks among the greatest encounters in physics history. A foundation established by the Belgian chemicals magnate Ernest Solvay paid for twenty-eight dapper men and one elegant woman to spend a week at a swank hotel and institute in Brussels giving lectures about quantum theory (which were later published) and having informal chats (which were off-the-record to keep everyone loose). Einstein didn't give a talk of his own, but made his case against the Copenhagen Interpretation during a question-and-answer period.

His argument was an updated version of the bubble paradox. A quantum wavefunction will spread out over space like an inflating bubble, yet the particle it represents will turn up in just one specific place when you go to look for it. What pops the bubble? What keeps the particle from appearing in more than one spot? Something must orchestrate the collapse to ensure that the particle materializes in one place and one place only. Yet no force is operating in this scenario—neither electricity nor magnetism nor gravitation. Nor *could* a force be operating, because the effect occurs instantaneously over a potentially infinite distance. The effect must be nonlocal; it must clash with the theory of relativity. Einstein put it this way at the Solvay meeting: "The probability that *this* particle is found at a given point assumes an entirely peculiar mechanism of action at a distance, which prevents the wave continuously distributed in space from producing an action at *two* places." This action at a distance, he told his colleagues, "implies to my mind a contradiction with the postulate of relativity."

To Einstein, the natural conclusion was that there wasn't any bubble that popped and left a particle behind. Instead, the particle was already sitting, waiting, at its observed location. No nonlocal orchestration is needed. This possibility sometimes goes by the name of "realism," because the particle *really* had a position all along, even if quantum theory failed to tell you where it was. The particle's position is what physicists call a "hidden variable"—"hidden" in the sense of not appearing in Schrödinger's or Heisenberg's equations. A more complete theory would include such a variable. Einstein had been trying to concoct such a theory based on his earlier ideas about a guiding field, and the French physicist Louis de Broglie presented such a model at the Solvay meeting. "I think Mr. de Broglie is right to search in this direction," Einstein said.

Bohr implicitly accepted the nonlocal versus incomplete dilemma. He agreed that the equations predicted nonlocality, telling Einstein: "The whole foundation for causal spacetime description is taken away by quantum theory." But Bohr couched his response in terms of what we can and can't measure, as opposed to what physical process could coordinate particle behavior over vast distances and what problems such a process would pose. He thought it enough that quantum mechanics provided "some mathematical methods which are adequate for the description of our experiments." What more do you want, people?

The two men spent several days deep in discussion over breakfast and dinner. We will never know for sure what they said, but all reports indicate that their conversation veered off from nonlocality and revolved instead around randomness—specifically, Heisenberg's uncertainty principle, which quantifies the degree of randomness inherent in particle behavior. Einstein repeatedly attempted to evade the principle and Bohr blocked him every time, leaving the impression that he had bested the frizzy-haired genius. One witness, the Austrian-Dutch theorist Paul Ehrenfest, memorably wrote: "Like a game of chess. Einstein all the time with new examples ... to break the UNCERTAINTY RELATION. Bohr from out of philosophical smoke clouds constantly searching for the tools to crush one example after the other. Einstein like a jack-in-the-box: jumping out fresh every morning." Meanwhile, Bohr punted on Einstein's central concern about links between distant locations in space. With this missed connection began the misunderstandings that would hinder the acceptance of nonlocality for half a century.

The 1930 Debate

By the next Solvay meeting, three years later, Einstein had devised another scenario to make his point. It was the first draft of the argument that all later debates would focus on. Suppose you have a box filled with photons rattling around like Tic-Tacs in their plastic container. One escapes through a hole and flies off into the cosmos. The box with the rest of the photons recoils in the opposite direction. Because the entire system of particles is described by a single wavefunction, the fates of box and escapee remain linked. Some time later, you measure the position of the box, from which you can calculate the photon's position.

This means one of two things. Either measuring the box does something to the photon or it does nothing to the photon. The Copenhagen Interpretation chooses the first option. It says that before you measure the box, both it and the photon are in limbo, having no specific position. After you measure, the wavefunction collapses and the photon pops up somewhere. Whatever instrument you used to measure the box must be acting like a remote control. Press the button and— boom! The particle instantly condenses from a vague haze of potentiality into a real pulse of light. Because the photon is moving at the speed of light, the remote-control signal would have to go faster than light to catch up to it. "Were that kind of a physical effect from 'B' on the fleeing light quantum to occur," Einstein later wrote, "it would be an action at a distance, that propagates with superluminal velocity. Such an assumption is of course logically possible, but it is so very repugnant to my physical instinct." According to the second option—measuring the box leaves the photon untouched—the particle already exists at the location where you find it, even if the wavefunction is blind to this fact. The Copenhagen Interpretation is misleading you into thinking you have a remote control when, alas, you do not.

In short, this new scenario posed the same dilemma as before: quantum theory is either nonlocal or incomplete. Unfortunately, one feature of Einstein's presentation in 1930 sowed perennial confusion, even among his sympathizers. Einstein noted that, instead of measuring the box's position, you could measure its momentum. It's as if your remote control has two buttons on it, one that causes the photon to materialize with a definite position, another that causes it to materialize with a definite momentum. But this extra functionality is secondary. The main issue is that you have a remote control at all.

Years later, Einstein tried to clarify this point: "That which really exists in B should therefore not depend on *what kind* of measurement is carried out in part of space A; it should also be independent of *whether* or not any measurement at all is carried out in space A [my italics]." A predecessor of mine at *Scientific American* who worked with Einstein on an article once told me that the great man didn't take too kindly to editing. But if an editor had prevailed upon him to cut out the first clause of that last sentence and leave only the part after the semicolon, the world would have been a wiser place.

Partly as a consequence of this confusion, the conversation between Einstein and Bohr got sidetracked for a second Solvay meeting in a row. Bohr fixated on the red herring of choice. He assumed Einstein was claiming that you could ascertain both a precise position and a precise momentum for the photon at the same time, which Heisenberg's uncertainty principle specifically rules out. If entangled particles violated the principle, quantum mechanics would not merely be incomplete, but wrong. Lore has it that the great Dane stayed up half the night to analyze the measurement procedure and, in the morning, triumphantly declared that the uncertainty principle held and quantum mechanics was saved. But according to others involved in the discussion, Einstein no longer had any quarrel with the uncertainty principle. He accepted that precise position and momentum measurements were mutually exclusive, and he considered quantum mechanics a logically consistent theory. His target was the Copenhagen Interpretation, whose proponents failed to own up to the nonlocality their view implied.

The EPR Paper

There would be no more great Solvay debates. In 1933 the Nazis raided Einstein's lakeside house, mobs of students burned his books, and few German physicists stood up for him. He renounced his German citizenship and left Europe, never to return. Two years later, ensconced at the Institute for Advanced Study in Princeton, he finally committed his worries about nonlocality to print. The paper universally known as EPR—for Einstein and his two younger coauthors, Boris Podolsky and Nathan Rosen—ignited a fiery debate among the scattered physicists of the world, conducted by post, with as many as three letters a day flying back and forth. Schrödinger, who also had fled Germany, was so moved by the paper that he followed up with several of his own. One of his papers coined a name for the phenomenon that Einstein had identified: entanglement. Another put forward his famous morbid scenario of a cat that is both alive and dead at once, an illustration of how quantum limbo wouldn't be restricted just to tiny particles, but would infect things of all sizes, even ones with fur.

Einstein never liked how the EPR paper turned out and reacted like any other academic in this situation: he blamed his coauthors. The main trouble was the final section, added by Podolsky, which overreached Einstein's intended point and attempted a takedown of Heisenberg's uncertainty principle, thereby perpetuating the confusion that had derailed the 1930 Solvay debates. The following year, Einstein published his own version, which kept the spotlight on the dilemma between nonlocality and incompleteness.

Yet the damage was done. Bohr latched on to Podolsky's iffy additions and came out with a rebuttal that was widely taken as settling the case. When physicists try to explain what Bohr's rebuttal said in any detail, though, they sound like contestants on Monty Python's TV game-show parody "Summarize Proust," bumblingly trying to sum up *Remembrance of Things Past* in fifteen seconds. One wrote: "Most physicists (including me) accept that Bohr won the debate, although like most physicists I am hard pressed to put in words just how it was done."

To this day, Copenhagen remains the dominant interpretation of quantum mechanics. Bohr's victory had much to do with social factors. Whereas Einstein was, by his own admission, a loner, Bohr was a father figure who inspired intense loyalty. He and his acolytes wrote nearly all of the early accounts of the debates. They naturally stressed Bohr's contributions and played down Einstein's, which had the effect of playing down nonlocality, because that was Einstein's concern. In their retelling, nonlocality was elevated to importance only through a "misunderstanding" on Einstein's part. A member of Bohr's institute sneered: "I must grant him that if a student in one of their earlier semesters had raised such objections, I would have considered him quite intelligent and promising." Such condescension set the tone of debate for much of the twentieth century. Only in the past two decades have historians such as Arthur Fine dug through correspondence from the period, showed the daring of Einstein's thinking, and restored his good name.

Often in the history of physics, physicists couldn't make progress because they lacked the technology or background knowledge. Faraday, for example, could hardly have built motors and electromagnets before the battery had been invented. But quantum physicists had no such excuse. They could have grasped nonlocality just as easily in the 1930s as in the 1960s. In fact, they very nearly did. Historians have envisioned entirely plausible counterfactual scenarios in which, had events taken a slightly different turn, Einstein would have won over his colleagues. A strange confluence of personality conflicts and self-reinforcing misunderstandings got in the way, and it took a new generation of physicists, led by John Stewart Bell, to rebel against the widespread denialism.

4

The Great Debate

Early in 2011, I was eating pasta with the philosopher Tim Maudlin when he told me about an upcoming symposium in Dresden, Germany, that promised to be a major debate on quantum nonlocality. Maudlin thinks the quantum realm is nonlocal. Other people who had been invited to the event think nonlocality is a fallacy. I got the impression that the organizers had simply intended to bring together the luminaries of physics and only later realized that it was like asking both Democrats and Republicans to an election night party. "There may be a huge bloody argument," Maudlin said. "The organizers are afraid of that. They're telling us to be polite." Naturally, I hoped the participants *wouldn't* be polite, and I booked my airline ticket as soon as I got home.

It's not that I like smackdowns, but I'd become frustrated that researchers had failed to resolve their disagreements over the foundations of quantum theory. They've debated the existence of nonlocal influences since Bohr and Einstein first went at it in the 1920s, and by rights the matter should have been settled already. When I started looking into the question of whether nature is truly nonlocal or merely puts on a good show—and therefore whether our conventional notions of space are in as much trouble as it seems—I figured I'd attend a conference or two, have a few chats over coffee, and sort it all out. My plan got off to a good start. The first person I talked to made total sense.

The second person I talked to made total sense. So did the third. Unfortunately, they were all telling me completely contradictory things. Clearly the plan wasn't working. I wasn't so naïve as to think that a roomful of professors would agree, but at least I thought I should be able to pinpoint exactly *where* they disagreed—to boil the dispute down to a choice between equally defensible assumptions. Often I could. But sometimes when I tried to grasp the nub of the disagreement, I found myself clutching at vapors.

Even the debaters told me they were perplexed. Many of them have been friends for decades, but still feel a disconnect over this issue. When I asked them to explain why their opponents feel the way they do, they stared off into the distance for a few moments and threw up their hands. Sometimes their frustration boiled over. One skeptic accused nonlocality proponents of "laziness" and called their arguments an intellectual "morass." Another complained: "These people have a very high opinion of themselves." "He's just going like a battering ram," agreed a third. For their part, proponents thought the skeptics were guilty of boneheaded mistakes. "He's just plain utterly absolutely unassailably appallingly wrong," groaned one. Seeking a neutral(ish) party to guide me through this physics road rage, I approached the philosopher and historian Arthur Fine. The first thing he told me was: "Welcome to the Hobbesian world of foundations—a war of all against all!"

You might think that if scientists quarrel over such a deep question as nonlocality, they might make a point of getting together and hashing it out. Yet the debate is remarkable for how little debating there really is. In my conversations, I often got the impression that people were exchanging their views through *me*, like estranged spouses using their kids to convey messages to each other. The Dresden meeting, I hoped, would finally flush these exchanges into the open. Big surprise: that didn't happen. The conference attendees didn't come to any agreement. They gave their talks, we all went to a fancy restaurant for dinner afterward, and we chitchatted about politics and other comparatively polite topics. I hadn't flown all that way for nothing, though. I began to realize that the serial failure to reach consensus is fascinating in its own right, a very human response to the depth of the mysteries posed by quantum mechanics. And even if the protagonists never shake hands, the debate can reach closure in other ways. As we'll see, the opposing positions lead us to very similar conclusions about the fundamental unreality of space.

The differing attitudes toward quantum physics reflect the conflicting emotional impulses of science: to revel in mystery and to puncture foolishness. "Some folks just can't stand the thought that maybe fundamental physics—quantum mechanics—can't be reconciled with common sense," Fine says. "And other folks just can't stand the thought that maybe quantum mechanics *can* be reconciled with common sense! These are differences in temperament that are not likely to be settled in discussions over coffee. They are sometimes made to appear as philosophical differences . . . but if you look, although you certainly find the expression of philosophical prejudices, there are hardly any serious arguments about philosophy as such. Instead, philosophical language provides a kind of code for emotional differences." Scientists are never unbiased observers, nor should they be, because who would put up with the travails of a life in research if they weren't driven? If poetry is emotion recollected in tranquility, then science is tranquility recollected in emotion: a struggle to think carefully in the swirl of passionate curiosity.

The Case for Nonlocality

By the standards of physics, the argument for quantum nonlocality is as simple as they come. Einstein's original paper with Boris Podolsky and Nathan Rosen in 1935 was just four pages long; John Bell's follow-up thirty years later, six. Neither involves heavy-duty math. If anything, equations tend to get in the way—Einstein complained afterward that his point in the EPR paper was "smothered by the formalism." These two papers represent two distinct steps in the logic. Einstein posed a dilemma: quantum mechanics is either nonlocal or incomplete. Bell closed off the second possibility: he showed that not even incompleteness could avoid nonlocality.

To see what this means, let's go back to the quantum experiment I described in chapter 1. In this experiment, particles act like magic coins that you can use to perform a variety of tricks. Like ordinary coins, they turn up either heads or tails at random when you flip them. Unlike ordinary coins, their tosses can exhibit a peculiar pattern. In the simplest case, you have a pair of these coins and give one to a friend. The two of you flip your respective coins and they land on the same side every time: both heads or both tails. By Einstein's reasoning, there are two possible explanations for their synchrony. They might be cheat coins, with the outcome fixed in advance; for instance, you and your friend might be flipping identical double-sided coins. This is the incomplete fork of the dilemma—"incomplete" in the sense that an onlooker has only partial knowledge of the coins and thinks they're fair when they're really a scam. Or they might truly be magic coins, linked by some mysterious connection (the nonlocal fork).

Einstein leaned toward the incompleteness option. He and Louis de Broglie proposed that particles match because an invisible guiding field is herding them like a sheepdog. The particles always exist in certain positions, and a measurement simply reveals where they are at any given moment, just as flipping a double-sided coin reveals a preordained outcome. This mechanism, Einstein thought, could produce the appearance of nonlocality. But what seemed like a good explanation in the abstract never worked when Einstein tried to create a mathematical theory of it. At one point, Einstein wrote a paper, sent it to a journal for publication, and belatedly realized the theory was nonlocal—but by then, the article had already gone to the printers and Einstein had to call the editor to stop the presses. (You can do that sort of thing when you're Einstein.) As Bell showed, there was a simple reason why Einstein had such trouble avoiding nonlocality: nonlocality is unavoidable.

Bell used a tactic familiar to party poopers everywhere: challenge the self-proclaimed magician to perform a feat that unequivocally requires magic. In one variant of his test, you and your friend flip your respective coins with either your right hand or your left hand, and Bell demands that the pair of coins sometimes land on the same sides, sometimes on *opposite* sides. You have a total of four permutations: both of you flip with your right hand, both with your left, you with your right and your friend his left, and vice versa. In three of these four cases, Bell wants the coins to land on the same side. But in the fourth, he insists that the coins land on opposite sides. It doesn't matter which case is the odd man out, but for the sake of argument, suppose it happens when you use your left and your friend his right. The pattern depends on what both of you are doing, so the situation is inextricably nonlocal. Not even the trickiest trick coin could fix the outcome in advance.

For instance, you and your friend could flip identical double-sided coins, as before; that way, your outcomes will always be the same, which satisfies Bell's challenge 75 percent of the time. You fail the challenge—and expose yourself as charlatans—whenever your coins were supposed to land on opposite sides. Some alternative trick might produce the desired outcome in those cases, but only at the price of doing worse on the others. My colleagues and I once created a video to dramatize how the test might work in practice.°

Bell's scheme isn't arbitrary; it corresponds to specific polarizer settings in the experimental apparatus. Quantum coins meet the challenge about 85 percent of the time. The additional 10 percent is the benefit of nonlocality. (The fact that the magic is imperfect—85 percent

| YOU | | YOUR FRIEND | | |
Choice of Hand	Result of Coin Toss	Choice of Hand	Result of Coin Toss (for cheat coins)	Result of Coin Toss (for nonlocal coins)
left	heads	left	heads	heads
left	tails	right	tails	**heads**
left	tails	left	tails	tails
right	heads	right	heads	heads
right	tails	right	tails	tails
left	tails	right	tails	**heads**
right	heads	left	heads	heads
right	heads	left	heads	heads

4.1. Test to distinguish true nonlocality from fakery. You and a friend flip coins with either your left or right hands, chosen at random. If the outcome has been fixed in advance—for instance, using cheat coins—the choice of hand won't matter. But if the coins are acting nonlocally, the choice can have an effect. In this example, the results in boldface mark deviations that indicate nonlocality is operating.

° The video is posted on the book's website, spookyactionbook.com.

rather than 100 percent—is an interesting clue to the nature of quantum nonlocality, more on which later.) And this situation isn't an outlier. Physicists have discovered dozens of coin-like quantum systems that aren't explicable by any conceivable sleight of hand. They comprise two, three, four, billions—any number of particles.

Einstein argued that a deeper level of reality was the only hope for saving locality. Bell dashed this hope. Having established that nature is nonlocal, Bell wondered how the nonlocality might operate. He reasoned that spooky action requires a spook—some immaterial entity that conveys influences from one place to another. And there was a candidate for one: the guiding field. Although Einstein and de Broglie had proposed the guiding field as a way to *avoid* nonlocality, the American theorist David Bohm reimagined it in the early 1950s as a mechanism to *generate* nonlocality. This field is roughly like the gravitational field as conceived by Newton or the Force in *Star Wars*. By banging on the field in one place, you could rattle a particle anywhere else in the universe. In principle, such a field is capable not just of creating dainty patterns in polarized light but of punching your enemy in the face from halfway around the world, although in practice this would require you to track and manipulate individual particles with impossibly high precision. (Some theorists have imagined circumstances under which it might be possible, though, such as the extreme conditions of the big bang.) Few physicists took to Bohm's proposal, and even today most shun it, complaining that the guiding field is nonlocal. But that's the whole point of it. If, like Einstein and Bell, you think that quantum mechanics is nonlocal, then Bohm's proposal has the virtue of bringing this nonlocality into the open rather than attempting to conceal it.

Motivated to Disagree

The logic of Einstein's and Bell's arguments is hard to dispute. So why do so many people dispute it? I think their misgivings spring from three sources.

First, some researchers just don't like nonlocality. It conflicts with so many other aspects of science that their gut tells them it's wrong even if their brain can't tell them why. Skeptics pass judgment on nonlocality and then seek some logical argument to back themselves up. Psychologists have a name for this common human tendency: "motivated reasoning." Such judgment calls may seem woolly and dogmatic to those who are on the receiving end, but they are essential to science. Sociologists have found that the most creative scientists tend to be the most stubborn. It's a myth that scientists are supposed to be open-minded; if they were, every new idea that comes along would blow them around like a weathervane. What is a scientist's job, if not to build up a stable, coherent view of how the world works? Accordingly, researchers judge all ideas by how they fit into the larger framework of knowledge. They are dubious about anything that doesn't conform, however strong the case for it may seem at first. They have seen plenty of supposedly airtight arguments spring a leak.

A second ground for skepticism is that quantum-coin experiments smack of skullduggery. When you create the coins—for instance, when the laser beam triggers the optical crystal at the heart of Galvez's entanglement experiment to emit a pair of photons—you create them specifically to match each other. That sounds an awful lot like distributing a pair of cheat coins. Bell's party-pooping challenge appears to rule out such trickery: no conceivable cheat coins could stay matched under the wide range of conditions that quantum coins do. But skeptics are not entirely reassured. What about *inconceivable* cheat coins? Is the metaphor of a coin misleading us? Might particles have ways to stay in sync without communicating nonlocally?

Another suspect moment occurs when comparing the outcomes. You ask your friend, "What side did your coin land on?" He tells you. You exclaim: "Why, that's just what I got. What a remarkable coincidence!" Until you have this conversation, you can't draw any conclusions about nonlocality. So you might wonder whether the conversation itself is part of the act. After all, the conversation is a quantum process, and skeptics reckon you can't be too careful when it comes to quantum physics. Bell's reasoning has the qualities of an iPhone: it looks so simple at first—just one button!—but open it up and behold the intricacy.

Third, and perhaps most important, the putative nonlocality strikes many people as kind of pathetic. The effect is completely imperceptible at the moment it is supposed to occur. All you see when you flip a quantum coin is the same random sequence of heads, tails, tails, heads, heads, tails that you see when you flip an ordinary coin. A pattern is buried in there, but to see it, you need to compare this sequence with the sequence your friend observed. The sequence is like an encrypted message you can't read until your friend sends you the code key. And the only way he can do that is by some run-of-the-mill communications system such as e-mail, phone, or marathon runner. This extra requirement of performing a comparison makes entangled particles useless for transmitting a signal.

Physicists have tried and tried to find a way to evade this requirement, but nature has foiled their every attempt. The closest anyone has come was the physicist Nick Herbert in the early 1980s. In effect, his plan was to conduct a series of repeated experiments. Suppose you and your friend decide on a code: left if by land, right if by sea. Your friend flips his coin with either his left or right hand and leaves it lying there. You flip your coin, then flip it again, and again, sometimes with your left hand, sometimes your right, watching for patterns. If the outcomes of your left-handed tosses follow the laws of chance, while the right-handed tosses are skewed toward, say, heads, you can conclude that your friend flipped with his right hand and his coin landed on heads. Your friend will have succeeded in sending you a message: the Redcoats are arriving by sea.

Automate this process, add some futuristic packaging, and you'd have the "subspace radio" or "hyperwave relay" that science-fiction writers have dreamed of for real-time interstellar communication. A message would zip from transmitter to receiver faster than light. Herbert gave his scheme a suitably spacey acronym: FLASH, for "First Laser-Amplified Superluminal Hookup." At first even the great Richard Feynman could find no flaw in the system, but pretty much everyone knew there had to be *some* flaw, if only because faster-than-light communication would let you send a signal into the past and wreak chronological havoc—having your grandfather killed, and all that. In short order, theorists found it: quantum mechanics expressly forbids do-overs. Each pair of coins is one-time-use only; once you flip the coins, they break their connection, so you couldn't conduct those repeated experiments after all. Quantum mechanics gives you no sneaky back-channel for sending messages.

For skeptics, the impossibility of faster-than-light communications is very suspicious. If you see a pool of water in the desert but, try as you might, can't take a drink, you must be seeing a mirage. Likewise, if you see particles linked together, but can't use them to send a message, perhaps the supposed link is an illusion. Even someone you'd think would be sympathetic, Steve Giddings—he who has done so much to convince string theorists of nonlocality in black holes—says he doesn't consider quantum nonlocality the real deal: "EPR is not true nonlocality. You can't send a signal with it." Those who do think entanglement is true nonlocality acknowledge that they have, at the least, a public-relations problem. "The fact that one can't send signals in standard quantum mechanics is suggestive," Maudlin admits.

Driven by these general motivations, skeptics of nonlocality have suggested several escape clauses from Einstein's and Bell's reasoning. Some of these alternatives might seem dubious, but if you're going to question something as basic as locality, you might as well revisit all your other beliefs, no?

Option #1: Giving Up Free Will?

One option is that entanglement experiments were rigged at the big bang. Recall why nonlocality seems necessary: without it, the particles would have to be preprogrammed for every eventuality that might befall them, and vanilla quantum mechanics provides no way to do that. In the coin metaphor, for example, the coins must be set up to respond one way if you flip with your right and another way if you flip with your left, and what kind of coin has that capability? If the coins can't be prepared in advance for every eventuality, they must have some nonlocal connection to keep them in sync.

But this conclusion doesn't follow if particles know beforehand what you're going to do to them. Then they can be ready for just that one eventuality. That would be possible if, for example, your choice of measurement is preordained. Think about it: Why would you choose your right hand to flip the coin? Maybe that hand was itching, or you once had a girlfriend whose name started with the letter r, or you had chosen left a couple of times and figured it was right's turn. Human beings make decisions for any number of reasons, not all of them entirely sensible. Those reasons follow from earlier events, and ultimately the chain of causes can be traced to the origin of the universe. Like a cosmic flash mob, the particles of the universe were put on a course to come together in just the right way at just the right time, priming your brain cells to choose your right hand.

Maybe the same factors that prompted you to choose your right hand also ensured that your friend chose his right and that both coins landed on the same side. Unbeknownst to you, the universe could be in the grip of a vast conspiracy to harmonize your choices with the coin tosses, making you think it's nonlocal when it's really local. This idea goes by the name of "superdeterminism." What makes it weird is the connection between phenomena that by rights should be completely dissociated. What has the tumbling of a coin through the air got to do with a complicated deliberative process inside your skull?

Superdeterminism is often described as taking away our free will. Actually, it's much worse than that. Even regular determinism—without the "super"—has caused many people to doubt we have free will. Through the laws of physics, you can ascribe every choice you make to the arrangement of matter at the dawn of time. That doesn't necessarily mean your will is unfree: freedom can be an emergent property, one that particles do not possess, but assemblages of particles do. As far as you're concerned, your choices can be entirely open until you make them. Nonetheless, philosophers, scientists, and students talking late into the night still debate the question. Superdeterminism adds a twist of the knife. Not only is everything you do decided in advance, the universe reaches into your brain and stops you from doing the very experiment that would reveal its true nature. The universe isn't just set up in advance. It's set up in advance to fool you.

I wasn't even going to include this conspiracy theory in my list—it sounds too much like the plot of a Dan Brown novel—until I got talking to the Nobel laureate physicist Gerard 't Hooft, one of the fathers of the Standard Model of particle physics. He thinks locality is so vital that physicists need to explore even crazy-sounding ideas to preserve it. "I tend to adhere as much as possible to a conventional picture of space and time because I attach great importance to locality," he tells me. "Without locality I think that basic laws of physics will be very hard or maybe impossible to formulate." 'T Hooft points out that one man's conspiracy is another's law of physics. Lots of things in the world seem conspiratorial at first glance, but are the result of well-established principles whereby objects coordinate their behavior. The fact that the Moon spins on its axis at exactly the same rate it orbits Earth (thereby keeping the same face to us, or nearly so) is the work not of a cabal, but of laws such as the conservation of angular momentum. Likewise, some new law of physics could conceivably harmonize particles' properties with humans' measurement choices. "What looks like a conspiracy today may be due to a conservation law we don't know about today," he has explained.

Arthur Fine also speaks up for superdeterminism. What makes it plausible, he thinks, is the fact that the synchrony of entangled particles is very subtle. To produce it, a grand cosmic conspiracy wouldn't need to manipulate you like a puppeteer, just nudge you gently. If in principle you have 1,000 choices of which measurement to make, but the universe allows you to make only 950 of them, that slight restriction on your freedom would be enough to produce the illusion of spooky action. "Constraints are common in physics and it turns out that even a small amount of constraint is sufficient," Fine says.

Option #2: Particles Are Crystal Balls

A second way that particles might be ready for you is that they can see the future. They could be like the Nazi doctor protagonist of Martin Amis's novel *Time's Arrow*, who experiences his life in reverse, from death to birth. Particles' past could be your future. They could be shaped by events that, to us, have yet to occur. They could come into the world already "remembering" what is going to happen—specifically, the settings of polarizers they will later encounter—and be prepared to respond accordingly.

Proponents of this idea don't deny that entanglement is magical. They just think the magic is a type of precognition rather than a type of telekinesis. Whether this counts as an improvement depends on your point of view. Still, when physicists as renowned as Feynman and John Wheeler suggest that particles are precognitive, you've got to take the idea seriously. Even Maudlin accepts this as a logical possibility: "Reverse causation can, in principle, account for the phenomena." By merging space and time, Einstein's theory of relativity made it natural to think of moments of time as laid out like points in space, all equally real, even if our puny brains perceive only one instant at a time. The future should be able to influence the present as surely as the past does, and indeed the real mystery is why we don't routinely see it doing so.

You can think of reverse causation as a form of time travel. Physicists normally flinch from time travel for fear of causal paradoxes. In this case, though, no paradox can arise because the particles are unable to convey a signal, let alone a human traveler. The time travel is confined to the interval between a particle's creation and its measurement, which is enough to give particles a glimpse of their future, but not to open a portal into the past. You can't use entanglement to rewrite history or check tomorrow's stock prices.

However solid the case for reverse causation may be, you know you're in trouble when people say that time travel may be the lesser of two mysteries. With these first two alternatives to nonlocality, you can already see where things are going. Einstein's and Bell's arguments may not strictly be proof of spooky action at a distance, but they *are* proof that we live in a very weird world.

Option #3: Parallel Universes

A third alternative supposes that nonlocality is an illusion caused by the existence of parallel universes that we can't see directly. It sounds like a breathtaking leap. How could a few glowing digits on a lab readout convince you that we cohabit space with an infinity of other worlds?

In fact, the reasoning is straightforward. Start by rephrasing Einstein's argument in a slightly different way. Entanglement means that two or more particles are set up to match each other. They are commanded to match. Thou shalt match! says the theory. But what a hopelessly vague directive. It doesn't tell the particles *how* to match. If they're acting as coins, they need to be told which side to land on, and without this information, they enter a state of limbo, both heads and tails at once. The situation is like a scene from the old military comedy TV show *F Troop*. An army sergeant orders the troops: Everyone, march in the same direction! But he neglects to say *which* direction. Some troops head to the right, others to the left, and mayhem ensues. In the absence of further instructions, the only way the coins or troops can escape their limbo in a consistent way is to communicate with each other telepathically.

But there's a loophole in this argument. What if quantum coins don't ever snap out of their limbo? What if each coin lands on *both* sides? In that case, no nonlocal influences need operate. Needless to say, such a prospect conflicts with the evidence of our senses. Whenever you flip a coin, you see it landing on either heads or tails (or neither, if it stands on edge), but never both. But maybe you shouldn't trust your eyes. Maybe you perceive only one outcome even though both occur. This could happen if you, too, have gone into a state of limbo—an ambiguous condition of seeing the coin heads up *and* seeing the coin tails up. After all, a human observer is a quantum object as surely as a particle is. If a particle can be in limbo, then why can't you? This situation is similar to Schrödinger's infamous both-alive-and-dead cat, except that we're now imagining what it would be like to be that cat.

The key is that you never perceive this ambiguity in yourself, but only in others. In effect, you have multiple personalities that are oblivious to one another. One personality sees the coin turn up heads; the other sees tails. Going back to the *F Troop* metaphor, although the sergeant's confusing command produces mayhem, he'll never notice if he has a brain disorder worthy of Oliver Sacks. His left brain sees soldiers marching to the right, his right brain sees soldiers heading off to the left, and both hemispheres are pleased that the troops have obeyed the command.

This gives us a new way to interpret what happens when you and your friend flip a pair of quantum coins. When you ask your friend what he saw, each of your multiple personalities hears a different answer. The personality that saw heads communicates with the personality in your friend's mind that saw heads. The personality that saw tails communicates with the personality in your friend's mind that saw tails. Both selves concur that the coins were coordinated. Curiously, they reach this conclusion even though the outcome of the toss remains ambiguous.

Proponents of this view argue that split-mindedness is the natural way to interpret quantum mechanics. It has nothing to do with our brains per se, but is a product of how the universe is put together. The best-known account, known as the many-worlds interpretation, supposes that those other possible selves represent near-duplicates of you that exist in parallel universes. In our universe, you may see the coin land heads up. In another, "you"—that is, a creature who is not the same collection of atoms as you, but looks exactly like you and is utterly convinced it is you—sees it land tails up. Everything that can happen does happen somewhere in the plurality of worlds, although we have direct access only to the world we live in. We think that a given event has a unique outcome because we can't see the other worlds where the alternatives transpire.

Many worlds aren't the only way to account for split minds. Other physicists have proposed interpretations that amount to the same thing minus the cosmological prolificacy. The main point is that we have access to only one portion of reality. Different observers can reach seemingly incompatible conclusions about reality, and that's fine, because their conclusions apply only to the portion of reality they see. There's no need for nonlocal influences to enforce consistency. However impeccable the logic may be, the conclusion still sounds fantastical. To throw your entire sense of self (let alone the nature of the universe) into question strikes many people as a high price to pay for explaining some esoteric physics experiment. But such are the lengths to which physicists think they must go to avoid spooky action at a distance.

Option #4: Don't Be Realistic

The surreal quality of the above possibilities explains the appeal of a fourth and final option: that it's all just a big mistake—that Einstein's and Bell's arguments have been misconstrued and, on closer inspection, don't imply nonlocality or any comparably momentous phenomenon. This argument was the focus of that meeting I attended in Dresden. It was the annual conference of the German Physical Society, the world's largest professional society for physicists, and a veritable physics fest. In a tent village set up on the university campus, you could stroll around with a beer mug, shopping for power amplifiers and cryogenic vats. Inside the lecture halls, physicists gave talks with such titles as "Honey, I Shrunk the Laser." There was even an "Einstein Slam," where the theory of relativity met open-mic night. The highlight, though, was the afternoon devoted to the nature of quantum reality. People jammed a thousand-seat auditorium, sitting in the aisles, standing in the doorways, probably breaking the local fire codes, craning their necks for a view of the speakers.

Defending the fourth option was Anton Zeilinger, the Viennese experimental physicist who has done as much as anyone to turn quantum entanglement from theoretical oddity to practical reality. He tells me that in the Austrian village where he grew up, in the foothills of the Alps, he had a reputation for curiosity—it was curious how curious he was. "To this day, the villagers say they thought I was kind of crazy," he says. "I was sitting in the window of my kitchen looking out for hours." Zeilinger and Maudlin are kindred spirits in many ways—two rebels against the prevailing dismissal of Einstein's probing questions. "When I started physics in the 1960s, the attitude was, let's not worry about the fundamental questions," Zeilinger recalls. He learned quantum mechanics on his own rather than from classes, and never fell under the impression that past generations had already answered the deep questions. "I realized these guys didn't know what it means," he says. "Something was missing." Alongside his physics, he read widely in philosophy: Kant, Mach, Popper, Wittgenstein.

Where Zeilinger diverges from Maudlin is that he never saw the Copenhagen Interpretation as something to revolt against. To the contrary, he rather likes it. Being an experimentalist, Zeilinger naturally identifies with Bohr's and Heisenberg's high esteem for the act of making a measurement—and, in particular, with the Copenhagen thesis that measurements actively help to create reality. He objects to Einstein's realist view that measurements are passive operations that record what is already out there.

Zeilinger sees Bell's argument in this light: as disproof not of locality per se, but of "local realism," which combines Einstein's twin intuitions that physics is local (particles are mutually independent) and realist (particles possess specific properties in advance of being measured). If Bell's argument pertains to local realism rather than just locality, then only one of these intuitions is vulnerable. By analogy, if we're told that someone is not a feminist bank teller, we'd need more information to know whether the person is not a feminist, not a teller, or neither a feminist nor a teller. And in that case, maybe we should be getting rid of realism rather than locality. Zeilinger and his team have conducted experiments that are damning for realism.

In one, particles remained in contact with one another, so that locality was never in question. This was a pure test of realism, and particles flunked it. If you try to explain the measurement results by supposing that the particles had specific properties all along, you arrive at a contradiction. Translating to the coin metaphor, imagine that five people are sitting around a table. Each tosses a coin and compares the outcome to the person on her right. Because five is an odd number, at least one person will get the same result as her neighbor's. But that's true only for ordinary coins. For quantum coins, you could find that no two people get the same result, which is impossible if the flips have definite outcomes that precede the act of comparison.

To be sure, such experiments don't rule out realism altogether. Some realist theories, such as Bohm's guiding-field concept, remain viable. But the results are still suggestive. Only when you try to imagine what particles must have been like before a measurement do you have to invoke nonlocal influences. Stop trying to imagine them in that way and maybe you can preserve locality. "In my eyes, locality is not the problem," Zeilinger told the Dresden audience. "In my eyes, the main culprit is the idea of realism."

Confrontation in Dresden

When Zeilinger sat down, Maudlin stood up. "You'll hear something different in my account of these things," he began. Zeilinger, he said, was missing Bell's point. Bell did take down local realism, but that was only the second half of his argument for nonlocality. The first half was Einstein's original dilemma. By his logic, realism is the fork of the dilemma you're forced to take if you want to avoid nonlocality. "Einstein did not assume realism," Maudlin said. "He *derived* it." Put simply, Einstein ruled out local antirealism, Bell ruled out local realism, so whether or not physics is realist, it must be nonlocal.

The beauty of this reasoning, Maudlin said, is that it makes the contentious subject of realism a red herring. As authority, Maudlin cited Bell himself, who bemoaned a tendency to see his work as a verdict on realism and eventually felt compelled to rederive his theorem without ever mentioning the word "realism" or one of its synonyms. It doesn't matter whether experiments create reality or merely capture it, whether quantum mechanics is the final word in physics or merely the prelude to a deeper theory, or whether reality is composed of particles or something else entirely. Just do the experiment, note the pattern, and ask yourself whether there's any way to explain it locally. Under the appropriate circumstances, there isn't. Nonlocality is an empirical fact, full stop, Maudlin said.

The only way to avoid the inference of nonlocality is to doubt the legitimacy of experimentation, which is essentially what the above options do. Namely, it might be impossible to perform a truly controlled experiment (because of superdeterminism or reverse causation) or to record the experimental results fully (because of our inability to see other universes). But such options are anathema to an experimentalist such as Zeilinger.

Clearly, the fourth alternative to nonlocality falls into a different category than the first three. Few deny that predestination, precognition, and parallel universes are logical possibilities, but debate still swirls over whether the denial of realism is a valid option. And even if it is, what then? How do those distant particles stay in sync? What's behind the magic trick? Zeilinger and others who advocate antirealism are never very specific about what's going on. To the contrary, they deny any need for such an explanation. This is where a chasm of mutual incomprehension opens up.

•

When Maudlin ended, Zeilinger raised his hand. Here was the moment I had flown halfway around the world for. Would Zeilinger put his finger on some flaw in Maudlin's reasoning? Would he concede the point? Would deep thoughts crackle through the air like lightning bolts? In retrospect, I should hardly have been surprised when Zeilinger merely reasserted his conclusion: "This inference of nonlocality seems to be based on a rather realist interpretation of information. If you don't assume this, you don't need nonlocality." So much for the fearsome clash the conference organizers had panicked about.

Later on, David Albert, a Columbia University philosopher who is a close friend of Maudlin's, remarked on how the debate was eluding closure yet again: "Students should take note of an appalling fact: we're forty-five years after the publication of Bell's paper, and you're hearing today very deep disagreement about what was proved in that paper. Tim and I hold that it was nonlocality. Others say there's a choice between realism or locality. I would encourage others to take note of this disagreement and be anxious to get to the bottom of it themselves." Both Zeilinger and Albert individually tell me they feel dissatisfied with their inability to connect. Zeilinger says he's so bothered that he has hired a team member specifically to catalog the diverse interpretations of quantum mechanics and their presumptions. Albert has suggested that I organize an open debate. He says the public has an interest in sorting it out. "You're not getting your money's worth," he muses. I have duly tried to get Zeilinger and Albert together, without success.

Perhaps I shouldn't be surprised at these repeated failures to work things out. Einstein's and Bell's arguments may be compelling, but so is the principle of locality. Another philosopher whose advice I've sought, Jon Jarrett of the University of Illinois at Chicago, tells me, "Given the spectacular success of the worldview challenged by the results of Bell-type experiments, we ought to allow a great deal of room for reasonable and, in this case, many very brilliant people to disagree."

Being Pragmatic

Where does this all leave us? For the record, I think Maudlin is right about nonlocality. Einstein's and Bell's logic is just too compelling. But I have also developed a healthy respect for all the ins and outs, and I have lost my earlier confidence that logic alone can settle the case. Making sense of disputes in science is seldom easy, anyway. Comparing two positions is a process of subtraction, of finding the difference, and subtraction is an operation that amplifies errors. It's like the accounting in *David Copperfield*: "Annual income twenty pounds, annual expenditure nineteen pounds nineteen and six, result happiness. Annual income twenty pounds, annual expenditure twenty pounds ought and six, result misery." A change in expenses of less than 1 percent leads to a complete reversal of fortune. Likewise, a slight misunderstanding of one side of a debate can completely mislead you when evaluating which is right.

This doesn't mean we have to go around in circles, though. Zeilinger tells me an instructive story. While a college student, he and a cousin backpacked around France on motor scooters. One night they pulled up to a youth hostel recommended by their guidebook, only to find it closed and abandoned. They decided to sneak in and sleep there anyway. As they settled in for the night, Zeilinger opened a window for air. A short while later, his cousin got up to close it. A while later, Zeilinger reopened it. And so it went all night. "I was convinced it was stuffy; he was convinced it was drafty," Zeilinger remembers. Only in the morning did they notice that the window didn't have any glass in it. Their nocturnal passive-aggressiveness had been pointless. For Zeilinger, the moral is that you always need to question your own assumptions.

I wouldn't go so far as to say that the debate over nonlocality is as empty as that window frame, but the way it goes on and on suggests it's time to take a step back and gain some perspective. There are three ways to do that: one proposed resolution might gain practical acceptance simply because it's more useful than the others; disputed concepts might gain credence because they're part of a broader pattern rather than just an isolated curiosity; and different options might have common features that guide us forward regardless of which proves right. Let's go through these one by one.

Skeptics doubt nonlocality in large part because they think it's pathetic, on account of being unable to transmit a signal. Yet the people offering this complaint tend to be theorists. I've spent a good deal of time with experimentalists in their labs and not once have I heard them belittle nonlocality. To the contrary, they bubble over with excitement about what it can do.

For instance, although entangled particles can't transmit a signal on their own, they can be combined with ordinary light or radio signals to cram extra information into a message. In one experiment, Zeilinger and his colleagues sent a message using a stream of photons. Ordinarily, each photon can carry one computer bit. But by using a stockpile of entangled photons and a procedure called "dense coding," the physicists were able to pack *two* bits onto each photon they sent. In effect, some of the information passed by way of the nonlocal link.

Entangled particles also have the power to thwart black-ops government agencies trying to listen in on your communications. The reason is that each pair of entangled particles is strictly one-time-use only. Using a string of these particles, you can send your friend a code key. Once your friend reads the key, the entanglement is broken, like the self-destructing reel-to-reel tape on *Mission: Impossible*. Then your friend can send you an encrypted message in complete assurance of its privacy. If an eavesdropper gets to the particles first, he or she can't help but break the entanglement, and you'll know you're being watched.

Of course, it does little good to shield your personal data from the government if you willingly type it in on a website. Fortunately, entangled particles can help with that, too. In a procedure known as "secure" or "blind" computation, you can share data with a web server while keeping the contents strictly private. For instance, suppose you want to open a credit line and the server asks for your salary to make sure you qualify. You can send your salary figure in a scrambled form and prove you earn more than the threshold without ever disclosing the precise amount.

Every experimentalist I have talked to calls these phenomena nonlocal. Even Zeilinger does, despite his theoretical qualms. Technically, "nonlocality" for an experimentalist simply means whatever it is that distinguishes quantum particles from classical atoms. It might not be true nonlocality, but an illusion produced by, say, the existence of parallel universes. But most experimentalists still envision spooky influences zapping across space. It's just easier to think that way.

In short, nonlocality is winning practical approval just as Newton's law of gravity once did. Newton cautioned that the nonlocality implied by his law might be illusory, but a new generation of physicists grew up with the law, saw how powerful it was, and felt no need to explain it away. They accepted gravity as a force generated at a distance. As the nineteenth-century psychologist William James remarked, "As a rule we disbelieve all facts and theories for which we have no use." But scientists will quickly accept even the most outlandish fact if you show them what they can do with it. A pioneering quantum experimentalist, Nicholas Gisin of the University of Geneva, remarks that his students are growing up with nonlocality and take it for granted. "Young guys find it fascinating, but are not totally amazed," Gisin says. "The kids here say, that's just the way it is."

Thinking Beyond the Quantum

Besides its usefulness, nonlocality also stands above the other proposed explanations because it fits into a larger context. In the next chapter, I'll talk about instances of nonlocality elsewhere in fundamental physics, but even sticking to quantum mechanics, consider the peculiar fact that entangled particles can achieve a desired pattern of outcomes 85 percent of the time rather than 100 percent. The seeming arbitrariness of the number suggests to many that quantum mechanics is just one of a class of conceivable nonlocal theories. Nonlocality is not an all-or-nothing phenomenon, but a spectrum of possibilities.

Some types of nonlocality are weaker than Einstein's and Bell's variety. In the same paper where he coined the term "entanglement," Schrödinger described how an entangled particle can "steer" its partner—an extremely subtle type of remote control that doesn't even create a special pattern in a coin-type experiment. Since then, physicists have found other ways that nonlocality can remain muted. Even such an enervated nonlocality is enough to perform useful tasks such as evading government surveillance.

At the other end of the spectrum, physicists have imagined types of nonlocality that are extra-powerful. In an influential paper in the 1990s, Sandu Popescu of the University of Bristol and Daniel Rohrlich of Ben Gurion University in Israel envisioned "superquantum" coins that would achieve the desired pattern more than 85 percent of the time. As far as anyone can tell, superquantum particles don't exist, but they've been a helpful what-if scenario. Engineers would be able to use such particles to build devices that are otherwise impossible. For instance, consider one of the minor irritations of daily life: finding an available time for a meeting. It is a deceptively hard problem, and computers (even those that harness the power of quantum particles) are reduced to flipping through attendees' calendars and comparing them one day at a time. But a computer using *super*quantum particles could find an available time slot in a single step.

Conference planners and speed daters would surely love to have some of those superquantum particles. But the power would come at a price. A world in which such particles existed could suffer from a deep logistical inconsistency or a tendency for complex structures to dissolve. In some inscrutable way, human life would have been impossible if it were too easy to arrange meeting dates. By trying to pinpoint how much nonlocality is too much, Popescu and Rohrlich think they might learn why quantum mechanics is the way it is. In these ways, the concept of nonlocality demonstrates the intellectual fertility that physicists value so highly.

Nonlocality by Any Other Name

Perhaps the most remarkable fact about the nonlocality debate is that the various alternatives aren't as distinct as their proponents make them out to be. Spooky action. Radical predestination. Precognition. Parallel universes. Unreality. What a list! Whichever is right, physicists have won the jackpot. All these options—even those that supposedly preserve locality—suggest a layer of reality beyond our everyday notions of space. All interconnect particles and observers on the opposite side of a lab bench, of Earth, or of the known universe. The only difference is how.

To see this, let's go through the options again. The most straight-forward is Bohm's guiding field. In follow-ups to his original work, Bohm proposed that the field should be thought of as an all-pervading fluid. Waves can sweep through it like ripples across a pond, transmitting influences from one particle to another. In truth, this idea is little more than a placeholder for ignorance. Waves in this field are nothing like ordinary water waves; mathematically they ripple through a higher-dimensional abstract space, and do so at infinite speed. The infinite speed is not an idealization, but an essential element of this model. If the putative waves traveled at a high but still finite speed, interactions among groups of particles would transmit signals faster than light, and nothing like that has ever been seen.

As soon as you start uttering the words "infinite speed," you know something can't be right. Motion that is infinitely fast scarcely deserves to be called motion; the thing that is "moving" is already at its destination, so how can it be said to move there? The guiding field is not really a medium through which waves can propagate, but a mathematical restatement of the fact that every particle depends on every other particle. Bohm himself came to think that the guiding field isn't a real thing out there in the world, but a flashing warning light that spatial concepts have failed. Instead of thinking of space as being filled with a guiding field or fluid, he suggested that you shouldn't be thinking of space at all.

The next option, superdeterminism, is said to eliminate nonlocality. On closer inspection, it does nothing of the kind. It merely displaces nonlocality from the present day to the big bang. Some law of nature had to determine what the universe was like at time 0. This law was like an overbearing mother at her daughter's wedding, insisting that everything be just so. Particle number 1,618,034 and particle number 137,035,999 were given matching properties and set in motion so that 13.8 billion years later, after countless collisions and interactions, they would come together in a laboratory on Earth, a place that didn't even exist yet. In effect, the law had the entire evolution of the universe, including interconnections among particles, already built into it. How is that any different from letting the universe play out and act nonlocally on the fly? Even 't Hooft, the leading advocate of superdeterminism, is unsure whether it really restores locality. "That is indeed a very difficult question that I'm asking myself all the time . . . I still see the EPR/Bell experiment as a problem," he says.

The third alternative, reverse causation, interconnects particles by means of signals passing backwards in time. This scheme has an appealing elegance. It requires no problematic guiding field to convey the influence from one particle to another; the particles themselves carry it along their timelines by "remembering" what happens to them in the future. Because the influence travels with the particles, it never goes faster than light, let alone infinitely fast. So this setup obeys Einstein's theory of relativity, ensuring theoretical détente. Nor does reverse causation demand an extreme fine-tuning of matter at the big bang.

For all these advantages, there's an irony here. One of the main reasons physicists dislike nonlocality is that it might permit time travel. To avoid this problem, proponents of reverse causation propose . . . time travel. So what do we gain? It might be simpler just to accept the time travel that nonlocality would imply.

Indeed, one might argue that reverse causation isn't really an alternative to nonlocal interactions, but a *type* of nonlocal interaction. It amounts to a holistic view of reality. Nothing can be understood in isolation; you need to look at the larger system of which it is part, including conditions in both the future and the past. Einstein wouldn't have liked this any more than he liked spooky action at a distance. His prime objection to nonlocality was that it threatened to make the universe incomprehensible. He asked: How can you pinpoint the cause of an event when the cause might lie in some galaxy too far away to see? How can you conduct a controlled experiment when you can't isolate the experimental system from distant influences? And this same objection applies to reverse causation. How can you pinpoint the cause of an event when the cause might lie in the future? How can you conduct a controlled experiment when you can't isolate a system from events that are yet to happen?

The fourth alternative, parallel universes and other forms of split-mindedness, says that the synchrony among entangled particles is a kind of illusion, an artifact of the highly selective view of reality that we have by virtue of living in one universe among many. To mere mortals, the universe looks nonlocal, but to a god who is able to survey the full scene, it is strictly local. Like the other options for explaining entanglement, though, parallel universes are effectively nonlocal. Each universe must have internal coherence: the right personality in one observer must talk to the right personality in another observer. And this coherence is a breed of nonlocality. In addition, parallel universes pull the rug out from under experimental science in just the same way that overt nonlocality does. Experiments can never probe the full ensemble of universes where locality supposedly holds.

Parallel universes also have some rather disturbing implications. A crucial role of locality for Einstein was to define what it means to be an individual; without locality, separate things would lose their distinctness. Much the same happens in parallel-universe scenarios. In a large enough proliferation of universes, every conceivable arrangement of matter will repeat itself. More than one creature will answer to your name, share all your memories, and be equally convinced that it is you. Now *that* is spooky. Which one of them is you? You can't tell. You could be here, there, over there—in any number of places. So, if parallel universes preserve a notion of locality, it isn't fulfilling the functions that locality is supposed to.

The fifth and final alternative, denial of realism, is harder to pick apart. Its detractors think it's flat wrong, and its proponents expend more energy on talking down the other options than on laying out their own explanation for why entangled particles match. But let's give this option the benefit of the doubt. In effect, the antirealists are saying that no explanation is possible or needed. Arthur Fine (whose own views are not neatly categorized as either realist or antirealist) has a nice way of putting it. Lack of realism means before you look at a quantum coin, it's in a state of limbo, and the act of observing the coin causes it to snap to heads or tails at random. There's no reason that the coin snaps to heads rather than tails, or vice versa. It just does. Fine suggests taking this thinking a step further. Once you accept that single events have no cause, it's natural to suppose that *pairs* of events have no cause. If nothing causes a coin to snap to one side or the other, perhaps nothing causes its entangled partner to snap to the same side. "If the individual events are undetermined, then what, other than belief in determinism, would lead us to think that something 'must be' producing the correlations among them?" asks Fine.

In other words, the correlations that troubled Einstein and Bell may be no more mysterious than the patterns that show up in any sequence of random events. Toss a coin repeatedly, and it'll come up heads and tails in nearly equal proportions per the laws of chance. Fine proposes that quantum mechanics extends this principle. Toss a coin here and a coin there, and they could exhibit a pattern per a broader conception of the laws of chance. Fine calls these correlations "randomness in harmony." I have heard a similar description from Gisin, the Swiss experimentalist: "We have randomness, but this randomness can manifest itself at several locations."

By eschewing a mechanism for interconnecting particles, however, Fine implicitly makes a dramatic claim about the nature of those particles. When two people flip quantum coins, in effect they are flipping *one* coin, and the outcome appears in two places, like a light switch that turns on two lamps at once. So those seemingly distinct particles have lost their individuality as surely as they have in the other options for understanding entanglement. Dispensing with realism hasn't eliminated nonlocality, but merely put a different spin on it.

Compare Fine's randomness in harmony with Bohm's guiding field. Fine's interpretation flouts the principle of separability: two seemingly distinct events aren't really *two* events. Bohm's skewers the principle of local action: two particles can communicate infinitely fast. Therefore each of these possibilities obeys one aspect of locality and violates one aspect. Einstein thought a failure of either aspect "entirely unacceptable." Empirically, too, these options are deadlocked. Both forbid you from using nonlocality to transmit signals, and both do so for the same reason: you can't control the particles.

Throughout this book, I have followed Einstein and treated separability and local action as a single package. One reason is that, although you might distinguish between them in the abstract, they almost always go together in reality. For instance, suppose Bohm is right and the principle of local action doesn't hold. That means objects can act on one another even if they're not touching. Those objects have lost their independence. So, the violation of local action also deprives the principle of separability of any significance. Conversely, suppose Fine is right and the principle of separability doesn't hold. That means there are no individual objects. But if there are no individual objects, then how can you talk about objects making direct contact? The principle of local action becomes an empty statement.

Space Is Toast No Matter What

All in all, we don't need to resolve the nonlocality debate to know that space isn't performing the functions that people have always assumed it does. If one particle can influence another particle instantaneously, position becomes meaningless; to be anywhere is to be everywhere. If one event manifests in two locations, those two locations are not inter-connected so much as collapsed into one. Space is a hall of mirrors. What's the point of it, then?

There's a reason Einstein called quantum nonlocality "*spooky* action at a distance" rather than just "action at distance." It's profoundly different from the earlier types of nonlocality I talked about in chapter 2. Newton's force of gravity also acts instantaneously, but at least it weakens as you get farther away from an object, so it still has spatial qualities. Entanglement, though, not only operates instantaneously, but also loses none of its strength with distance. It's like the bond between lovers: it depends only on what went on between them in the past and loses none of its power no matter how far apart they may be. How romantic! But particles aren't supposed to be romantic. They're supposed to be gears in a spatial clockwork.

Instead of thinking of quantum nonlocality as an effect that operates within space, I think we need to take it as a sign that space itself is a doomed concept. "We should really be trying to understand these correlations as being due not to laws or influences, but as emerging in an organic way from an understanding of what the underlying degrees of freedom are," says Jenann Ismael, a philosopher at the University of Arizona. To adapt a saying by one of Einstein's teachers, individual objects by themselves fade away into mere shadows, and only a kind of union of them preserves an independent reality. The universe can't be cut up into separate regions of space, but must be considered holistically.

"Holism" is a word that has to be handled with care. Not only is it a slippery concept to begin with, it has become a banner for all sorts of discontent with science and modern life in general, and many scientists and philosophers react defensively by shunning the word altogether. I've seen researchers tie themselves up in knots trying to avoid describing quantum phenomena as holistic. I have no such qualms. Nonlocality *does* mean we live in a holistic universe, one that isn't reducible to its spatial parts. The world has qualities that are hidden to us when we view it piecemeal, but that reveal themselves when we take it all in at once. Ultimately that's why we can't use quantum nonlocality to send messages. To measure the holistic properties of an entangled pair of particles, you have to measure both particles jointly. Such properties aren't going to be apparent if you measure just one particle on its own. So you can't manipulate one particle in the expectation that someone watching the other particle will notice any difference.

The holism we're talking about here operates at a much deeper level than what yoga teachers and alternative-medicine practitioners speak of. Ironically, the sentiments that motivate these believers in holism—a feeling of interconnectedness with the rest of the natural world, and a frustration that modern medicine fails to attend to the whole person—depend on nonholistic physics. You can't feel interconnected if there is no "you." Your individuality and your integrity as a person require nature to consist of independent parts. So nature may well be intrinsically holistic, but it did not become capable of sustaining life until it partitioned itself. It is the *division* of the world, not its interconnectedness, that should awaken our wonder.

"Quantum correlations just happen, somehow from outside spacetime," Gisin concludes. To explain these correlations, physicists and philosophers will have to go beyond spacetime—and also beyond quantum mechanics. The theory lays out the options, but offers no definitive resolution, and the standard battery of experiments can tell you only so much. "It is tempting to think that the holism indicates a fundamental failing in our very notions of space and time," Maudlin has written. "After all, if . . . particles 1 and 2 are so fundamentally interconnected, perhaps we are mistaken to think that they are really distinct particles, or occupy different regions of space-time. And perhaps our notions of space and time could do with a fundamental revision. But the failure of reductionism in quantum mechanics does not force this option on us."

The situation changes when researchers look laterally to other domains of physics. The nonlocality doesn't go away, as many people hoped it would. It becomes even more strongly entrenched, and new types of nonlocality augment the particle synchrony that Einstein focused on. These new phenomena don't just whisper a failure of spatial concepts, but sing it loud. "I'm sure there is no story in spacetime," Gisin says, "but I'm sure there *is* a story."

5

Nonlocality and
the Unification of Physics

When Steve Giddings decided to climb Denali in 2003, he naturally chose one of the most dramatic and demanding ascents, the Cassin Ridge route. It crosses the Valley of Death, an avalanche and crevasse zone whose name is not merely metaphorical. Sometimes rescue teams can't safely recover the bodies; sometimes they can't even find the bodies. The Cassin is an all-around test of mountaineering skill: technical rock and ice climbing, crevasse crossing, Arctic camping, and, not least, patience. Giddings built up his endurance on the trails around Santa Barbara; he'd scramble up steep ridges with a fifty-pound backpack, return to the bottom, repeat. The Denali attempt started well. But his climbing buddy developed blisters so deep they threatened to cut into his Achilles tendon. Then a storm rolled in as they reached the advance base camp at fourteen thousand feet, holding them there. Even the standard route up was closed off. "We saw streams of climbers head up that route to a higher camp, at about seventeen thousand feet, get beaten back by the weather, and head down the mountain defeated," he recalls.

Giddings's efforts to study black holes had been the intellectual equivalent of his Denali expedition. These mysterious cosmic objects are an all-around test of modern physics. Because their gravity is intense, you need to analyze them using gravity theory—namely, Einstein's general theory of relativity. And because quantum effects are strong in them, you also need to consider quantum theory. So you can't get away with shirking one theory or the other to make the job easier. And when you do apply both theories to black holes, they don't mesh. To break the theoretical deadlock, Giddings proposed in the early 1990s that a nonlocal mechanism operates in these objects.

The idea stirred up a storm. In physics, as in mountaineering, it's hard to know when to press on and when to turn back; the field rewards stubbornness, but only up to a point. Giddings decided to move on to other problems. "Maybe the community wasn't quite ready for it," he supposes. To most of his colleagues, giving up locality didn't reconcile the two theories so much as forsake them. Einstein and the others who created these theories had done so with the express purpose of expunging nonlocality from physics. Newton's gravity acted at a distance, as if by magic, and general relativity snapped the wand in two. Likewise, quantum mechanics in its original incarnation covered how particles react to forces, but left out how those forces were transmitted; it implicitly assumed that forces leap across space. Physicists had to develop an advanced version called quantum field theory to fill in the mechanism of propagation. To this day, they present both general relativity and quantum field theory to their students and the public as the epitome of locality.

Yet physics theories, like mountains, have an uncanny capacity to surprise. In creating any theory, physicists cobble together various ideas gleaned from experiment and intuition. The result invariably transcends its original context. Heinrich Hertz, who helped to devise the theory of electromagnetism in the nineteenth century, remarked: "It is impossible to study this wonderful theory without feeling as if the mathematical equations had an independent life and an intelligence of their own, as if they were wiser than ourselves, indeed wiser than their discoverer, as if they gave forth more than he had put into them." So it was a century later with general relativity and quantum field theory. Whatever their creators' intentions may have been, the theories began to reveal a different side as physicists put them to use. The workings of the forces of nature turned out to be sparkling with nonlocal phenomena.

These instances of nonlocality, which are distinct from those that spooked Einstein about quantum mechanics, are the subject of this chapter. They indicate that the locality we observe in daily life may not be characteristic of the way things really are. Although forces act locally—their influence spreads through space at a limited speed—this locality doesn't seem to be rooted in the structure of nature. There are no separated entities that transmit and receive these forces; the world can't be partitioned into independent spatial pieces. And in that case space must not be the true venue of physics.

It has taken decades for most physicists to process these peculiar features of their theories. A tipping point came in the 1990s. At the time, theorists working on unifying physics were split into two main camps, both of which were reaching similar conclusions about nonlocality. Those who took the approach known as loop quantum gravity argued that force fields are a big tangle that binds together far-flung locations in space. Meanwhile, the competing approach of string theory grew so far beyond its original conception—that subatomic particles are vibrating stretches or coils of energy—that physicists came to refer to it as "the theory formerly known as strings." The string theorist Juan Maldacena, then at Harvard and now at the Institute for Advanced Study, developed a concept known as "AdS/CFT duality," in which locations that seem far apart can lie right on top of each other. What appears to be spatial distance is in fact a difference in energy.

Don Marolf, a colleague of Giddings's at UC Santa Barbara, and one of the few people who move in both loop-theorist and string-theorist circles, recalls how nonlocality emerged from obscurity: "The discussion increased significantly in the late 1980s and early 1990s due to the birth and growth of loop quantum gravity, which made the issue seem more pressing and relevant to a growing community. Then Maldacena's discovery of AdS/CFT in 1997 made the issue of great interest to a *much* larger community."

Consequently, when Giddings resumed arguing for nonlocality in black holes, in 2001, he enjoyed a very different reception than he had before. Nonlocality no longer seemed so crazy. To the contrary, it now began to strike theorists as entirely natural. If quantum field theory and gravity theory exhibited nonlocality, it seemed quite plausible that some nonlocal mechanism might operate in black holes. Having achieved this professional vindication, Giddings made his Denali climb two years later. For ten days, he and his limping buddy waited out the bad weather in the base camp. When the skies finally cleared, Giddings set out with another team of climbers and reached the summit. "The most sublime part of the trip was skiing down the lower Kahiltna Glacier the following night, late at night but still in the Alaska twilight, with shadows and colors playing across the nearby peaks of the Alaska Range," he says. "I'll never forget how beautiful that was."

Quantum Field Theory

As I talked about in chapter 3, Einstein and others developed quantum mechanics to resolve paradoxes in the classical conception of light. It was kind of a letdown, then, that the first mathematical formulations of the theory couldn't explain light. The equations nicely described material particles moving at modest speeds, but didn't impose any speed limit, as relativity theory demanded. So they couldn't handle things moving near or at the speed of light. Considering that light moves at the speed of light, this was a serious limitation.

Quantum field theory was the relativity-friendly, hence light-friendly, sequel to quantum mechanics. To develop it, physicists in the 1920s and '30s took two approaches, depending on whether they thought light is ultimately particle or wave. Some, such as Paul Dirac in England and, later, Richard Feynman in the United States, were partial to particles. They adopted the atomist picture of little billiard balls caroming around. You just had to make the picture more sophisticated by supposing that the balls can be created and annihilated on the fly. That way, atoms can emit light by creating a photon and absorb light by destroying one. Classical electromagnetic waves are built up from great big effervescent gobs of photons. Other electromagnetic phenomena, such as static electric and magnetic forces, can be thought of as photonic billiards, too. Although the theory originally described just photons and electrons, it was later extended to neutrinos, quarks, Higgs particles, and the rest of the subatomic zoo.

Other theorists, such as the Austrian-born Wolfgang Pauli, gave primacy to waves. For them, the world is like a pond in a rainstorm, trembling with ripples that form, spread, and merge. The "pond" is the electromagnetic field that invisibly fills the space all around us. Waves of every sort course through it: long, short, tall, squat. Theorists apply quantum mechanics to each individual type of wave and sum up the whole glorious mess. In this view, what we perceive as "particles" are not motes of matter, but units of wave energy.

In a remarkable case of convergent evolution, the particle and wave viewpoints led to the same equations. There was no need to choose between them. You could think of light—and not just light, but all forms of energy and matter—as either particulate or wavy, and you could call the subject either theoretical particle physics or quantum field theory. (Even now physicists use these terms interchangeably.)

A big part of the achievement was figuring out how relativity theory fits in. The founders of quantum field theory didn't incorporate *all* of relativity theory; they left Einstein's explanation of gravity for another day and concentrated on ensuring that particles or waves wouldn't outrun light. This was surprisingly tricky. As Pauli pointed out, a moving particle or wave might cease to exist at any moment and bequeath its energy to new particles or waves, confounding your attempts to follow it, let alone check whether it's breaking the speed limit. Instead of attempting to impose a limit on speed per se, Pauli focused on the *consequences* of speed—like a cop who writes you a speeding ticket based not on a radar gun reading, but on the fact that you got home before you were supposed to.

He came up with a rule called "microcausality," which puts a basic restriction on cause and effect. It chops space into two parts: one that light can reach within a certain time, and one that it can't. If you send a signal and the intended recipient is inside the first region, he should get it; if he's in the second region, you're out of luck. Crucially, it doesn't matter exactly what is conveying the signal: a wave, a particle, or something else entirely. All you care about is the effect, or lack thereof, on the receiver. There's some controversy over whether microcausality is the essential lesson to draw from relativity theory, but the alternatives also divvy up space into zones that light can either reach or not.

All these features of quantum field theory reflected physicists' intuitions about locality. The two competing worldviews from which the theory emerged—particles or waves—are both local. Particles are localized scraps of matter and interact with one another only by making direct contact or by enlisting other particles as middlemen. Waves in a field convey forces from one place to another in a continuous sweep without any miraculous nonlocal jumps. Indeed, the whole reason that Michael Faraday and James Clerk Maxwell introduced the electric and magnetic fields in the nineteenth century, as I discussed in chapter 2, was to guarantee locality. Microcausality or an equivalent rule ensures that particles or waves move at a finite speed, providing a measure of isolation among separate regions of space.

To say that physicists "developed" quantum field theory implies that they knew what they were doing. All along, though, they were stumbling around and often deeply doubtful about what they'd gotten themselves into. By combining elements of both quantum mechanics and relativity theory, physicists had performed a shotgun marriage with unforeseeable consequences. To this day, physicists struggle to fathom what quantum field theory is telling them about the world. As we saw in the previous chapter, ordinary quantum mechanics isn't a model of transparency, either, but at least it's fairly simple mathematically; you don't even need calculus to do useful calculations. Quantum field theory is another story. It has a deserved reputation as the most mathematically badass subject in the sciences. Even experts hang on for dear life. Joe Polchinski at the Kavli Institute for Theoretical Physics in Santa Barbara tells me he retook his first course on the theory twice over and got his Ph.D. still not feeling entirely at ease with it.

It's the job of philosophers to make sense of physics, but the fearsomeness of quantum field theory has scared most of them off. One of the few who isn't intimidated is Hans Halvorson of Princeton University. To the contrary, he likes nothing more than hacking his way through mathematical thickets. He's the kind of person who does his own taxes every year and complains it's too easy. "When I started doing quantum field theory as a grad student, I was so happy," he recalls. "There's an endless supply of problems." His love of burrowing into algebraic formulae is just what the subject demands. If anything, he has the opposite problem from other philosophers: he struggles with conceptual thinking, the hallmark of philosophy. But the fact that philosophizing doesn't come naturally to Halvorson appeals to him: it is his mountain to climb. "I tend to make things too mathematical," he says. "A mathematician here told me, 'Step away from the equations for a minute.' A *mathematician!* . . . The hard part is interpreting what the mathematics means."

Parting with Particles

What it means, apparently, is that our world consists neither of particles nor of fields, at least not in the way these terms are commonly understood: as structures that embody the principle of locality. Physicists still speak of "quantum particles" and "quantum fields," but that's like saying "open secret" or "paid volunteer." The adjective "quantum" connotes "kind of like a particle, but not" or "has some decidedly unfield-like characteristics."

Take the particle viewpoint first. According to the older theory of quantum mechanics, a particle's position and velocity are uncertain. You don't know where it'll turn up or how fast it'll be moving. But at least it does turn up *somewhere*. That's no longer true once you factor in relativity theory, as you need to do if the velocities involved are a fair fraction of the speed of light. The reason goes back to a key fact about quantum uncertainty. The velocity and position of a particle are not independent quantities. If you know the spread of possible velocities, you can calculate the spread of possible positions using the Heisenberg uncertainty principle, and vice versa. Relativity theory spoils this conversion by requiring that the uncertainty principle be observer-independent. Now, when you translate velocity to position, you find that different positions are no longer mutually exclusive. You might find the same particle in two different places at once, or the particle in one place and its energy somewhere else. The combination of quantum mechanics with relativity violates locality in what was, for Einstein, its most basic sense: the stipulation that all things have a location.

For relativity to be implicated in nonlocality is a stunning reversal. Einstein's cosmic speed limit was supposed to have banished nonlocality, not entrenched it. "When we see this happen, we say, 'Oh great, we thought we had prevented this from happening,'" Halvorson remarks. "People think relativity is contrary to nonlocality, but here relativity *gives rise* to nonlocality."

In an influential analysis in 1949, the quantum theory pioneer Eugene Wigner and his student Ted Newton showed that a particle could have an unambiguous location only if relativity did not apply to location measurements. In that case, though, observers would no longer agree on what the universe looks like, introducing a worrisome subjectivity into physics. That's a steep price to pay, and it wouldn't even solve the problem. Now observers would disagree not only on what a particle's position is but on whether it even *had* a position. Some could narrow the particle's location to a confined region, while others could see it materialize anywhere in the universe. Those observers who did find a particle at some well-defined location might see it suddenly leap to the far side of the universe—an effect that, if real, would let engineers build a faster-than-light communications system. Okay, you might say, let's give up trying to pinpoint particles and simply try to count them. Even this modest ambition will fail, as different observers will come up with differing answers.

All in all, quantum field theory says that looking for particles is like playing a shell game. You can't get a fix on them, you see them disappear from one spot and reappear in another, you can't even agree on how many there are. The darned things are beginning to look like a scam. Most physicists and philosophers have concluded that little billiard balls just can't exist in our universe. "There isn't anything that's truly localized," Halvorson says.

The "particles" that appear in the equations of quantum field theory are actually a type of wave. Such "particles" exist at no one location, but throughout the entire field, just as a note picked on a guitar string doesn't exist at any particular place along the string, but spans the entire length. Their only claim to the term "particle" is that they represent discrete chunks of energy and momentum. And even this denuded usage of the word "particle" works only when energy and momentum can be divided into autonomous chunks. When fields are interacting intensely, waves are so jumbled that particles no longer exist, even by a liberal definition.

Physicists routinely talk about particles. Almost everything written about physics, from textbooks to bathroom graffiti, talks about particles. For many purposes, it's still fine to talk about particles. But there is no such word in the language nature itself speaks. Every time you see what you think is a particle, you need to look more closely. For instance, the most widely used formulation of quantum field theory—the system of stick-figure diagrams devised by Feynman that I talked about in chapter 1—shows particles as interacting at specific places in space and time. These diagrams are commonly used to study scenarios such as the collision of two particles. But any given scenario involves an infinity of diagrams, although you might get by with "just" hundreds or millions. An individual diagram doesn't represent anything out there in the real world; only the aggregation of diagrams is meaningful. So, the diagrams are just a useful mathematical device, a way to break down a big problem into manageable chunks, like saying that the average American family has 1.9 kids and 2.3 cars. The particles that appear in these diagrams, including the "virtual particles" that often figure into discussions of physics, are our mental constructs. "They can't actually be saying anything about reality," Halvorson says.

Experimental physicists, for their part, build particle detectors like the MacGyverish cloud chamber I duct-taped together in my basement, and it stands to reason that particle detectors detect particles. In fact, all they are really registering is little bursts of wave energy—transient disturbances like glints of sunlight off the choppy surface of a pond. It's tempting to connect the dots and assume that flecks of matter are flying through the instrument. Resist the temptation.

Farewell to Fields

If the world isn't made of particles, by default it must consist of fields, right? Maxwell visualized electric and magnetic fields as an assemblage of conveyor belts and rolling drums, like an Industrial Revolution steel mill. As children of the information age, we might compare fields to a flat-screen TV or computer display: a mosaic of little pixels. To carry the analogy further, each "pixel" is a little localized doodad, like a particle, but fixed in place. It has attributes analogous to brightness and color. It can exert a force on objects, respond to the forces that objects exert on it, and interact with its immediate neighbors. If, for instance, you put a compass in a magnetic field, nearby pixels will grab hold of the needle and twist it to align with their own direction. Conversely, if you wave a magnet, nearby pixels will respond, their neighbors will respond in turn, and a ripple will spread across the screen. The metaphor works well as long as you imagine the screen as having an infinite resolution, so that its elements are not tiny squares of light, but geometric points of zero size. They form a continuum without any seams or gaps.

That was the idea, at any rate. Yet quantum field theory subverts the classical notion of field as surely as it sweeps away particles. The namesake "fields" of quantum field theory are arrays not of things, but of operations. The field can act, and be acted upon, at specific locations. It can cause a compass needle to spin, and it can absorb energy that you pump in. But what's producing these effects? The theory doesn't say. It suffers from all the interpretive ambiguities we have already seen for the older theory of quantum mechanics. Quantum field theory specifies what a field does, but not what it is. And whatever the field may be, it definitely can't be an array of pixels. The earlier arguments against particles also rule out pixels sitting at each point in space. Nor can the field be any other kind of localized structure, because it has nonlocal properties. What the field does in one part of space depends on what it does elsewhere. The concept of the field, says Halvorson, "seems like it's backfired in quantum field theory. You can cut space into blocks, but the blocks are highly interconnected. We've undermined our own theoretical advance."

Entanglement on Steroids

Quantum field theory has two distinct types of nonlocality. The first is a supercharged form of quantum entanglement. Inklings of it first appeared in the work of Feynman, who showed that particles can break the Einsteinian speed limit as long as they don't get caught—that is, as long as they don't convey information or cause something to happen that otherwise wouldn't. Later physicists recognized Feynman's faster-than-light propagation as a version of the spooky synchrony of entangled particles. Instead of particles being entangled, points in the field are. If you put two measuring instruments at different locations in a field, their readings can match even though nothing passes through the space that separates them.

The effects of relativity theory make field entanglement more than just a warmed-over version of ordinary particle entanglement. Recall that any given observer sees space as divided into two pieces: one that is close enough to receive a signal in a given amount of time, and one that is too far away. To respect this division, the waves that ripple through the field must add up in just the right way. Their high degree of co-ordination implies a powerful bond between far-flung locations. "The very fact that stuff is constrained in momentum . . . means that stuff is connected in position space," Halvorson says.

The upshot is what he and his colleagues have termed "super-entanglement." Whereas regular entanglement connects certain aspects of two or more particles, such as their polarization, superentanglement links *every* aspect of *everything*. It's not just a particle here and a particle there. Every point in space is entangled with every other point. That even includes points lying beyond the observable universe. And the entanglement is as firmly knotted as a kindergartner's shoelaces. You can sever the bonds between entangled particles, but you can't ever disentangle a field.

This intricate web of nonlocal connections has impressive consequences. It lets you do the magic-coins experiment of chapter 1 without going through the rigmarole of creating and exchanging entangled particles. Just pluck the coins out of thin air—that is, from the fields that are already around you—and flip them. They will land on heads or tails at random, and always in harmony with each other. For dramatic effect, you could even do the experiment in a perfect vacuum—not a particle in sight. Just stick a pair of instruments into the field, they will pick up its residual random jittering, and the readings will match. "You may have forgotten to arrange the experiment," Halvorson says. "The universe will have arranged one for you."

How kind of it. But this charity has limits. The experiment is damned hard to do and, despite various proposals, no one has tried it for real. One reason is that, in order for the field to satisfy the micro-causality rule and related conditions, the correlations must be highly specific in location. The situation reminds me of the spotty cell-phone service in my house: I get five bars while standing in my kitchen, but zero when I step into the dining room. If I switched cell-phone providers, I might get a signal in the dining room, but none in the kitchen. By analogy, a physicist's measuring instruments might detect spooky correlations at first, but go blank if moved a smidge farther apart. The field will still be entangled at the new locations, but the experimenter would have to switch to another type of measuring instrument to continue detecting the correlations. The field has "got this nonlocality at all distances, but any given type of nonlocality will die off," Halvorson says. If physicists can solve this technical problem, though, fields may become the preferred source of entanglement for quantum cryptography and computation.

Field entanglement enables entirely new classes of phenomena. You can entangle an atom with a field, then entangle the field with a second atom. At that point, the two atoms will be entangled with each other even though they never interacted directly. Entanglement also orchestrates phases of matter besides the usual solid, liquid, and gas. Those ordinary phases are distinguished by how atoms and molecules are arranged, such as the neat rows of a crystal or the helter-skelter of a gas. The new phases are more elaborately ordered, with large-ensemble choreographies worthy of a Bollywood movie. In these situations, nonlocal effects don't get washed out with distance, and they endow materials with properties that once seemed magical, such as superconductivity, a condition in which electric current flows without resistance.

Until fairly recently, most theorists thought of these phenomena as mere parlor tricks. "Traditionally, my colleagues and I were not so impressed with entanglement," recalls Nima Arkani-Hamed, the ayatollah-escaping theorist from chapter 1. "Laypeople love it, but—big whoop! All the action is elsewhere. The fact that paying attention to entanglement might pay off in nontrivial ways is surprising." In fact, he now thinks the tangling-up of fields could be the defining feature of quantum field theory: "The right way to think about quantum field theory may be to think of mutual entanglement among regions."

Going back to our flat-screen TV metaphor, you might visualize field entanglement as a nest of wires crisscrossing the back of the screen to link pixels together. For most purposes, this isn't a bad image, and you can still think of the field in basically the same way that Michael Faraday and James Clerk Maxwell did. But deep down it fails. Entanglement doesn't mean that the brightness and color of one pixel can become coordinated with the brightness and color of other pixels. It means that individual pixels don't actually have brightness or color values; only groups of entangled pixels do. An entangled field has holistic qualities that do not exist in any one place, but span the entirety of space.

Gauge Crossing

The second category of nonlocality in quantum field theory isn't a quantum effect but a structure that is latent in the nature of electric, magnetic, and other force fields. In fact, you can catch a glimpse of it by looking outside your window to a high-voltage power line and marveling at the birds perching on it in complete safety. The birds don't fry because high voltage, per se, has no effect. Birds, sadly, do get electrocuted, but only when they touch two wires at once, creating a path for current to flow through their little bodies. What gives them the shock is the *difference* in voltage—or, to be more careful with terminology, the difference in the electric potential.

This feature of electricity is known as gauge invariance, so called because electrical effects don't vary if you raise or lower the potential, as long as you keep potential differences the same. Two wires could be at 0 volts and 120 volts, or 120 volts and 240 volts, or 1,000,000 volts and 1,000,120 volts, and you could never tell which. The most riveting demonstration of this principle is an experiment Faraday first performed in 1836 and which never fails to get a crowd oohing and aahing at science museums today. He built a twelve-foot wooden box, draped it in copper wire and tinfoil, and climbed in. His assistants electrified the outside using a giant electrostatic generator, giving the cube and all within it a high voltage relative to the rest of the room. Lightning bolts began to crackle all around as if the Gates of Hell had opened. On the inside, though, Faraday conducted electrical experiments as if he were in the quietest seaside cottage. The high voltage had no effect on him. He wrote: "I went into the cube and lived in it, and using lighted candles, electrometers, and all other tests of electrical states, I could not find the least influence upon them, or indication of anything particular given by them, though all the time the outside of the cube was powerfully charged, and large sparks and brushes were darting off from every part of its outer surface."

An analogous, if less stageworthy, situation also occurs for magnetism. There is a magnetic potential as well as an electric potential, and in fact you can trade off raising and lowering the two potentials, making the definition of voltage even more fluid. Forces other than electromagnetism are gauge-invariant, too. In fact, "gauge" is probably theoretical physicists' second favorite word after "coffee." Gauge invariance is central to their conception of how forces operate. But what does it mean? Why exactly does only the potential difference matter?

Theorists commonly take gauge invariance to be self-justifying— a symmetry, or pleasing harmoniousness, like that of a circle. Just as you can spin a circle without changing its appearance, you can raise or lower the potential without changing anything measurable. By this way of thinking, gauge invariance is an expression of the elegance of nature, and physicists are softies for elegance. But a growing number of theorists find this explanation unsatisfying. Raising or lowering the potential is not a physical manipulation of the world like spinning a circle; only the potential difference is ever under our control. It also seems rather suspicious that the supposed elegance leads to so many mathematical complications. "Textbooks rhapsodize about gauge symmetry," Arkani-Hamed has said. "Gauge symmetry is a complete fiction. It's all in our heads." He thinks gauge invariance is instead a sign of nonlocality: "You could take the existence of gauge fields as the first of many, many hints that we need to abandon locality."

Locality holds that each point in space has properties independent of other points, which suggests that the wire should have a potential defined in some absolute sense. Maybe it's 1,000,000 volts. If a bird standing on this wire plants its other claw on a wire at a potential of 1,000,120 volts, the potential difference will be 120 volts. This sounds very sensible. If you have a difference, surely you must have two numbers that you're subtracting. But if you ask an electrician to tell you the potential of a single wire—not the difference between it and another wire, but the potential of the wire on its own—he or she will give you a puzzled look (and then charge you $140 for his or her time). A theorist will be struck equally dumb if you ask how the potential varies from moment to moment. Although Maxwell's equations make a prediction for the value of the potential, they also let you add whatever fudge factor you want to that value. Locality has led us to assume the existence of something that's impossible to measure or to predict definitively, and that's no good.

Maxwell thought the particulars of a given situation—the setup of batteries, coils, and magnets that you wire together—would fix the value of the potential at every point in space. But he found himself predicting that disturbances in the potential would zap from one side of the universe to the other instantaneously. Recoiling from this blatant nonlocality, his immediate successors, such as Heinrich Hertz, deemed the very concept of the potential an "inanity" and vowed, somewhat melodramatically, to "murder" it. They rewrote Maxwell's equations to do away with the electric potential as a fundamental ingredient of nature. The equations we now call "Maxwell's equations" aren't really Maxwell's.

Whatever these equations should really be called, they say that all electromagnetic phenomena are produced by electric and magnetic force fields, which have a definite (and measurable) strength everywhere in space, in accordance with locality. The potential is derivative: it represents how much energy a force field can impart to charged particles. Mathematically, the potential is a useful but optional concept. If you really wanted to freak out that electrician, you could talk about electricity without ever uttering the word "volt." And if the potential doesn't really exist, it doesn't matter that you can't measure or predict it. The fact that you can raise or lower its value with impunity has no deeper significance, and the disturbances that propagate instantaneously are purely fictional.

•

So, Victorian-era physicists figured that their act of theoretical homicide had reconciled gauge invariance with locality. By the mid-twentieth century, though, theorists such as Paul Dirac realized that giving primacy to the force field didn't eliminate the mystery. The force field may look local, but on closer inspection it has some nonlocal properties of its own. The strength of the field at a given location is not free to take on any old value, but only a limited menu dictated by values elsewhere in space. The values are interdependent, or, in the argot, mutually "constrained." If the field is like a flat-screen TV, it is a broken flat-screen TV, whose pixels are short-circuited together so that one pixel can be turned on only if a certain combination of other pixels is also turned on. Such a screen can display only a limited range of images— like a TV that can show the Super Bowl, but not *Downton Abbey*.

An example of these constraints involves the orientation, or polarization, of light waves. Light can be polarized in two distinct directions. That's why modern 3-D movies use polarized light: the movie projector shines two images on the screen, one in each polarization, and the 3-D glasses ensure that each eye sees the polarization intended for it. But there's something a bit weird about this. Why aren't there *three* polarizations? After all, space has three dimensions—that is, three possible directions of motion. Some types of waves, such as seismic waves in the Earth, oscillate in all three. Why not light waves? Because of the constraint. It shackles together different points in the field, thereby restricting their range of motion like a line of can-can dancers locked arm in arm, kicking out in front, but never to the side. Mathematically, the constraint is defined in terms of the properties of individual points in space, yet it deprives those points of their autonomy. Part of what locality means is freedom from such constraints. Different locations in the field are supposed to be independent—able to affect one another by sending out ripples through space, but not rigidly locked together as a matter of logical necessity.

To deepen the challenge to the old sanguine view of gauge invariance, in 1959 the theorist David Bohm and his then-student Yakir Aharonov proposed a quantum version of Faraday's cage experiment. Electrons, rather than lighted candles, function as the test probes. This experiment is tricky to do and physicists didn't pull it off until 1985, but the analogous test for magnetism is easier and was done within a year of Bohm and Aharonov's proposal. Result: particles do behave differently inside a high-voltage cage. The wave pattern they create is shifted. It is shifted even though the electric force field inside the cage is zero and, by the standard rules of electromagnetism, not a thing should happen.

For most physicists, this came as a shock, and few believed it at first. But they came to accept that localized structures are mismatched to the reality of electromagnetism. The electric potential is too much: a given potential difference could arise from an infinite variety of absolute potential levels, creating a richer set of possibilities than nature ever realizes. The electric force field is too little: it isn't rich enough to capture the shifting of electron wave patterns, just as a computer screen can't do justice to the saturated colors of a Chagall painting. Theorists needed some structure that is just right. And with the local possibilities exhausted, this third bowl of porridge presumably has to be nonlocal.

•

It took even longer for philosophers to catch up to these developments. One of the first was Richard Healey of the University of Arizona. Whereas many people trace their love of science to some early influence—a parent or teacher who opened their eyes to the wonders of the universe—Healey says he was drawn by the *lack* of any such influence. "I was surrounded by people who knew all sorts of stuff, but they didn't know about science," he recalls. "That was my corner on the market." School reaffirmed his contrarian passions. "I suffered through high school physics," he says. "I hated it. I knew what these guys were hiding from me. I wanted to know more—I certainly wasn't learning that in high school." His career, too, has been built on watching what other people were doing and then doing something else. Because philosophers in the 1990s largely neglected gauge invariance, he naturally gravitated to it. "Back then, I felt very lonely," he once admitted to his colleagues.

What fascinates Healey is that gauge invariance could reveal a whole new type of nonlocality, completely distinct from the entanglement phenomena that Einstein and John Bell agonized over. The electrons in the Aharonov-Bohm experiment need not be entangled or otherwise specially prepared. They are serving more as spectators than as actors. By virtue of their wave nature, they are sensitive to features of the world that ordinary objects are oblivious to, and they bring out a kind of sleeper nonlocality within force fields.

Though different from entanglement, the nonlocality of gauge invariance does have comparable effects. "People draw distinctions between the Aharonov-Bohm effect and Bell inequalities, but they have an awful lot in common," Healey says. Entanglement binds particles together into an integral whole, with collective properties that individual particles lack. Gauge invariance likewise endows fields with features that exist in no particular place, but span a wide region of space. In both cases, the system is not merely the sum of its spatial parts—contrary to the aspect of locality that Einstein called the principle of separability. A field is separable almost by definition, so the failure of this principle is bad news for the concept of fields. At the same time, both entanglement and gauge invariance do respect the other aspect of locality that Einstein identified, the principle of local action, since neither phenomenon will let you transmit a signal, remote-pilot a drone, or beam happy thoughts to your loved ones far away.

So, if forces aren't conveyed by fields and aren't conveyed by particles, what *are* they conveyed by? Healey advocates an idea that goes back to Dirac and gained traction among physicists in the years after Aharonov and Bohm's experiment. It's based on a crucial clue: the potential at a single point may be ambiguous, but if you add up the potential at multiple points forming a closed loop, *that* quantity is unambiguous. For instance, you might begin at the back door of your house, walk in a big circle around your yard, and return to your starting point, measuring the electromagnetic potential along the way. Individual readings will depend on your arbitrary choice about what to define as zero volts, yet the sum of all your readings will be independent of this choice. Somehow this series of readings feels out a structure in nature that individual readings fail to pick up. Whatever causes electricity to flow and magnets to stick to a fridge, then, might not be a neat array like a computer screen, but an intricate lacework like a crocheted scarf. "Loops, not points, are the natural home of electromagnetism," Healey says.

The loops are spread out (violating separability), but transmit electric and magnetic forces by jiggling one another (satisfying local action). Under ordinary circumstances, you can't tell the difference between loops and classical fields. In effect, the loops are pulled so tight that they look like points in a regular grid. But the true nature of the fabric becomes unmistakable when one of the stitches gets caught on a nail and yanked out. In the Aharonov-Bohm experiment, the electrified cage acts like that nail, distending one of the loops and creating effects that Maxwell never dreamed of. When you try to capture these effects using a local structure such as the potential, you are left with some ambiguity.

The moral of the story is that quantum field theory is messing with our notions of space. It's not a theory of localized building blocks such as particles or pixels; in fact, such things seem to be impossible. Instead, it's a theory of delocalized structures, whether loops or something else. Technically, this fact doesn't demand that we give up our notions of space. We can still imagine those loops, or whatever they are, as existing within space. "We're not losing the points," Healey says. "It's the structure that's defined on them that's nonlocal." But situating loops within space is like holding a rock concert in a symphony hall. You can do it, but everything seems off. Physicists and philosophers have always inferred the nature of space from the behavior of matter. The ancient Greek atomists invented the concept of space in order to give particles a place to play. For modern theorists, space is the substrate of fields. If particles and fields don't really exist, space loses its raison d'être.

Einstein's Theory of Gravitation

Quantum field theory really pulled a practical joke on physicists. It is local in some ways, but nonlocal in others, and it makes us wonder whether space is all it's cracked up to be. A similar saga has played out for the other pillar of modern physics, general relativity. Arguably, the failure of space is even more striking in that theory.

One day in autumn, Don Marolf and I were talking about gravity while sitting in the student center of the UC Santa Barbara campus, eating salads and looking out over the lagoon. But hang on. How did I really know I was sitting in the UCSB student center on a certain day in autumn? The principle of locality says that I had a position, the student center had a position, and when these two positions coincided, I was there. The GPS coordinates on my phone matched those of the center, and the date matched the calendar on the wall. But this seemingly straightforward procedure doesn't stand up on examination. "To ask a question about here, we should know what we mean by 'here,' and that's not so easy to do," Marolf says.

One obvious complication is that California is tectonically active. The crustal plate on which Santa Barbara sits is moving northwest by a couple of inches per year relative to the rest of North America and to the national latitude and longitude grid. So the student center has no fixed position. If I come back some years from now and go to the same coordinates, I'll find myself sitting in that lagoon. Mapping companies must periodically resurvey tectonic zones to account for this motion.

You might suppose that the student center still has a position defined in an absolute sense by space itself. Yet space is no more stable than a tectonic plate. It can slide, heave, and buckle. When a massive body shifts, it sends tremors through the spacetime continuum, resculpting it. The position of the cafeteria might change as a result. This process, rather than Newton's mysterious action at a distance, is how gravity is communicated from one place to another according to Einstein's theory. Like geologic tremors, gravitational ripples propagate at a certain finite speed—namely, that of light.

To grasp the reshaping of spacetime, our minds have to overcome a hurdle of abstraction. Spacetime is not as tangible as a geologic landscape. We can't see it, let alone discern its shape. Yet we catch indirect glimpses. Objects that are moving freely through space, unhindered by other objects, are like raindrops streaking across a car windshield, revealing the curve of the glass: they trace out the shape of space. For instance, astronomers routinely observe rays of starlight that start off as parallel, pass near a giant lump of mass such as the Sun, and afterward intersect. Textbooks and articles describing this effect often say that the Sun's gravity has bent the light rays, but that's not quite right. The rays are as straight as straight can be. What the Sun has really done is to alter the rules of geometry—that is, to warp space—such that parallel lines can meet.

The morphing of space and time is not just the stuff of exotic physics. It governs the motion of any falling object. Baseballs, wineglasses, expensive smartphones: things that slip out of your hand accelerate toward the floor because Earth's mass warps time. (The warping of space plays only a minor role in these cases.) "Down" is defined by the direction in which time passes more slowly. Clocks at sea level tick more slowly than clocks on the summit of Denali; a watch strapped to your ankle will fall behind one on your wrist. In human terms, the deviations are small—parts in a trillion at most—but enough to account for the rate at which falling objects pick up speed. When you see an apple fall from a tree, you are watching it roll across the contours of time.

•

Although the shape-shiftiness of spacetime explains away the kind of nonlocality that Newton talked about, it creates several new varieties. For starters, general relativity allows spacetime to bend back on itself to create a tunnel, or wormhole, connecting two otherwise distant parts of the universe. The entrance of one would look like a doorway that leads you not to the next room, but anywhere else in the universe. You could step through to, say, Alpha Centauri, a journey that would take at least four years the long way. Astronomers have yet to discover any wormholes, and the type of matter needed to pry them open may not exist, but general relativity permits them, and any prediction of Einstein's has to be taken as a live possibility.

Strictly speaking, wormholes would fully respect the principle of locality. If you ducked through one, you'd follow a continuous path through spacetime, with no mysterious jumps, and at every point in your journey, you'd travel slower than light. All that would change is the length of your journey. But notice my weasel words "strictly speaking." Unstrictly speaking, wormholes tick off every box on our nonlocality checklist. To begin with, a wormhole is an inherently nonlocal structure. By definition, it spans multiple locations; it is not localized to one place. To say that spacetime has a wormhole is like saying that a pretzel has a knot: a statement about the overall shape, as opposed to any one part of it. Two universes—one with a wormhole, one without—could look identical at every place, and you'd notice that one had a wormhole only when you compared separated locations to see whether they were interconnected by multiple paths.

A wormhole's effects, too, are nonlocal. By means of the wormhole, two things that appear to be distant could in fact be adjacent or perhaps even the same thing seen from two different angles. Many theorists have speculated that a mini-wormhole could join entangled particles and explain why those particles give matching results when measured (an idea I will return to in the next chapter). To be sure, a wormhole's effects may not be obvious to you as nonlocal. If, for example, something passes through the wormhole, knocks into one of the particles in your body, and causes that particle to veer off its original course, all you'll notice is that something weird happened. You won't be able to tell the difference from any of the other random and unaccountable things that befall you every day. Some theorists have suggested this would explain why quantum processes have unpredictable outcomes and why the basic parameters of nature, such as the strength of forces, have such inexplicable values.

A wormhole also restricts the behavior of matter in the universe by creating a loop. A wave might ripple through space, enter a wormhole, return to its starting point, and overlap with itself like a snake eating its own tail. Consequently, the wormhole would alter wave undulations over a vast region and possibly prevent stable waves from ever forming. The weirdness quotient goes way up if the wormhole connects points separated not just in space but also in time—in other words, if the wormhole is a time portal. Now the wormhole creates a *time* loop, with all the potential for the logical paradoxes that science-fiction fans love to think about. To foil such paradoxes, time travelers must lack the freedom of action that people ordinarily have. For instance, suppose you go back in time and try to shoot Grandpa. Something will stop you. You know something has to, or you couldn't have made the trip to begin with. You might fumble the gun, the gun could misfire, a meteor could fall from the sky and knock the gun out of your hands, and so forth. No matter how often and how hard you try, the universe will conspire to spare your target. To you, it'll look like a very peculiar set of coincidences.

To explain those coincidences, it wouldn't be enough to scrutinize your immediate environs; you'd have to look to the overall structure of the universe. The requirement of logical consistency places a constraint on matter that passes through the wormhole (and, in some cases, on matter in other parts of the universe, too), and such a constraint is a form of nonlocality. What happens in one place depends on what happens elsewhere.

•

A second type of nonlocality in general relativity is even more elemental. It comes out of the theory's core innovation: that there's no such thing as a place outside spacetime, no external or absolute standard to judge it by. This seemingly self-evident proposition has remarkable consequences. It means that spacetime not only warps, but also loses many of the qualities we associate with it, including the ability to define locations.

Disavowing a god's-eye perspective, Don Marolf says, "is very subtle, and, honestly, Einstein didn't understand it for a long time." Previous conceptions of space, including Newton's and even Einstein's own earlier thinking, supposed that space had a fixed geometry, which would let you imagine rising above space and looking down on it. In fact, at one point, Einstein argued there *had* to be an absolute reference point or else the shape of space would become ambiguous. For a sense of why the ambiguity arises, consider how we experience geography in everyday life. We might suppose there is a unique "real" shape to the landscape—what Google Earth shows—but in practice the shape is defined by the experience of being embedded within that landscape, and that experience can vary. A student running late to his exam, an athlete hobbling on a sprained ankle, a professor walking with a colleague deep in conversation, and a cyclist yelling at all those pedestrians to get out of the way will perceive very different campuses. A short distance for one may seem an interminable crossing to another. When we eschew the view from on high, we can no longer make definitive statements about what is where.

In an epiphany in 1915, Einstein realized that the ambiguity is not a bug, but a feature. He noted that we never observe places to have absolute locations, anyway. Instead we assign positions based on how objects are arranged relative to one another, and—crucially—those relative locations are objective. Everyone wandering around the college campus will recognize the basic ordering of places. They'll juxtapose the UC student center with the lagoon rather than put them on opposite sides of campus. If the landscape buckled or flowed while preserving these relations, the denizens would never know. So it is for spacetime. Different observers may ascribe different locations to a place, but will agree on the relations that places bear to one another. These relations are what determine the events that occur. "If George and Don met in a certain café at noon in the first spacetime," Marolf tells me, "they would also do so in the reshuffled spacetime. It's just that in the first case this would have occurred at point 'B' and in the reshuffled case it occurs at point 'A.'"

The cafeteria, then, is situated at either A *or* B *or* C, D, E—an infinity of possible positions. When we say it's located at such-and-such a place, we're really using a shorthand for its relations to other landmarks. Lacking definitive coordinates, the cafeteria must be situated by the things within and around it. To locate it, you'd need to search the world over for a place where the tables, chairs, and salad bar are arranged just so and where a patio overlooks a lagoon bathed in the golden sunlight of Southern California. The position of the student center is a property not of the center, but of the entire system to which it belongs. "The question you asked in principle refers to the whole spacetime," Marolf says.

•

This ambiguity of position bears more than a passing resemblance to what we saw earlier with electromagnetism. It's gravity's version of gauge invariance. Not being able to tell the difference between points A, B, and C is like not being able to tell whether the electric potential is 0 volts, 120 volts, or 1,000,120 volts. For gravitation as for electromagnetism, the ambiguity of localized measurements is a form of nonlocality. And here it is happening not because of the properties of particles and fields, but because space itself is unable to support any localized structure. Points in space are indistinguishable and interchangeable. Because they lack any differentiating attributes, whatever the world consists of must not reside at points. Quantities such as energy can't be situated in any specific place, for the simple reason that there's no such thing as a specific place. You can no sooner pin down a position than you can plant a flag on the sea. Those quantities must be holistic—properties of spacetime in its entirety.

Furthermore, the multiple equivalent shapes of space are described by different configurations of the gravitational field. In one configuration, the field might exert a stronger force in one place than it would in another configuration, with compensating changes elsewhere to maintain the relative arrangement of objects. Points in the gravitational field must be interlinked with one another, so that they can flop around while collectively still producing the same internal arrangement of objects. These linkages violate the principle that individual locations in space have an autonomous existence. Marolf has put it this way: "Any theory of gravity is not a local field theory. Even classically there are important constraint equations. The field at *this* point in spacetime and the field at *this* point in spacetime are not independent."

None of this means space is a mere fiction. It still does have some independent reality—it can expand or contract, waves can propagate through it, and it can exist even when devoid of material objects—so it can't be entirely reduced to a set of relations among objects. Under most circumstances, you're entitled to think of spatial locations. You can designate some available chunk of matter as a reference point and use it to anchor a coordinate grid. You can, to the chagrin of Santa Barbarans, take L.A. as the center of the universe and define every other place with respect to it. In this framework, you can go about your business in blissful ignorance of space's fundamental inability to demarcate

locations. "Once you've done that, the physics looks like it's local," Don
Marolf says. "The dynamics of gravity is completely local. Things move
in a continuous way, limited by the speed of light." But the properties
of gravity are still only "pseudo-local." The nonlocality is always there,
lurking beneath the surface, waiting for its moment to emerge.

In short, Einstein's theory is nonlocal in a more subtle and insidi-
ous way than Newton's theory of gravity was. Newtonian gravity acted
at a distance, but at least it operated within a framework of absolute
space. Einsteinian gravity has no such element of wizardry; its effects
ripple through the universe at the speed of light. Yet it demolishes the
framework, confounding our intuitive picture of space as a kind of
container in which material objects reside. General relativity forces us
to search for an entirely new conception of space.

•

Black holes are the prime example of a place where the nonlocality
of gravity comes out of hiding. I say "place," but when it comes to black
holes, that word becomes problematic. A black hole is not a solid object; its perimeter or "horizon" is just a hypothetical line in space that
marks the point of no return for infalling objects. And where is that
line? It's very hard to tell. Suppose, God forbid, our Sun collapses to
form a black hole and a group of space tourists goes to take a closer
look. The tour company assures them they're safe as long as they stay at
least three kilometers from the center of the hole, which is the calculated radius of a black hole with the Sun's mass. But that is false adver-
tising. If the black hole sucks in more matter, it will enlarge and the
tourists could find themselves on the inside, unable ever to return
home. The location of the horizon depends not only on how strong the
hole's gravity is now, but how strong it *will be*. The black hole has time
on its side; freedom is never assured.

I once went to a lecture by Gary Gibbons, a colleague of Stephen
Hawking's at the University of Cambridge, in which he discussed this
strange feature of black holes. "In order to define this horizon, you need
to know what happens for all of time," he said. That, in turn, means you
need to know what happens for all of *space*. Horizons "are *highly*
nonlocal," Gibbons continued. "They're tricky in that respect. You can't

put your finger on this horizon; we may ourselves be inside one at this moment and we would never know." A black hole could be opening up around our planet like a sinkhole beneath a mining town, slowly and invisibly.

To deepen the puzzle, all the stuff that falls into the black hole piles up at the very center, its so-called singularity. General relativity theory says that matter reaches infinite density and spacetime rips open like an overloaded grocery bag. Where is the singularity located, then? The spacetime against which its position would be defined has ceased to exist. There is literally no there there. In a strange sense, a singularity exists nowhere and everywhere. It is not a localized object, but a holistic feature of spacetime.

•

Another vivid demonstration of gravitational nonlocality occurs if space has a boundary. For space to have a boundary sounds like a throwback to Aristotle, who thought the cosmos was encased in a crystal sphere like a ginormous snow globe. That idea went out with lace ruffs. As far as modern astronomers can tell, space stretches forever in every direction, with no end of galaxies. But this is a contingent fact rather than a hard-and-fast requirement. There's no law of physics that forbids the universe from having an edge. Physicists have toyed with hypothetical models in which one or more dimensions of space are finite in size.

Even an infinite universe can still have a boundary—one that is located at infinity, which, to a theoretical physicist, is as real a place as any. To see why, consider what it means *not* to have a boundary. Earth's surface is an example: you can buy an around-the-world airplane ticket, set off to the west, and return to your starting point without ever doubling back on your route. The universe could, in principle, have a spherical shape, so that you could venture out into space and circumnavigate the whole thing. By default, if you can't do a full loop, the universe must have a boundary. Even if you never encounter a solid wall or a dragon-filled abyss, the endless wastes of infinity act as an insurmountable obstacle. Infinity can have exactly the same structure as any other region of space. Objects can reside at infinity. Events transpire there. An infinitely distant boundary clearly poses practical difficulties for space travelers; only the starship of the imagination could ever reach it. Conceptually, though, it's just like the snow globe.

Whether finite or infinite, the boundary is the final frontier's final frontier, and like all frontiers, it is a place apart. Gravity may be king through the bulk of the universe, but its writ does not run to the edge of space. Out there, the gravitational field is immobilized because the boundary fixes the shape of space like a mold dictating the form of a clay pot. "Spacetime at the boundary does not fluctuate," Marolf says. "It is effectively nailed to the boundary, so there is no dynamical gravity." Because gravity doesn't operate along the boundary, neither does nonlocality (at least not of the gravitational sort). The boundary provides an absolute reference point, so any cafeterias and student centers that happen to exist there have objectively defined locations. Quantities such as energy are completely unambiguous. You could set up your measuring instruments on the boundary and take readings with surety.

Although spacetime is nailed down at the border, elsewhere it's as floppy as ever. So the universe has a peculiar dual character: part behaves locally (the boundary) and part (the bulk) nonlocally. Consequently, the holism of the universe, a notion that sounds almost mystical, becomes very tangible. Locality on the boundary enables you to make measurements there, and nonlocality in the bulk connects those measurements to the rest of space. "Quantities that one might a priori think are independent are in fact locked together," Marolf explains. "Thus one can have boundary observables that are identically equal to bulk observables." Because the boundary tracks what is happening throughout the cosmos, it's an image of the entire universe—a perfect image, with no loss of fidelity. An observer watching it would know everything you're doing.

This happens because, as we saw above, things are never truly localized. Slapping a boundary onto space has flushed this ambiguity into the open. You can have two identical things, one in the bulk and the other on boundary, and by all measures they're a single thing manifested in two locations, as if you saw a tail fin and a spout poke up above the sea and realized they were one whale. Interestingly, the boundary mirrors the bulk even though it spans fewer spatial dimensions. If dimensionality is so fungible—if the universe can be described

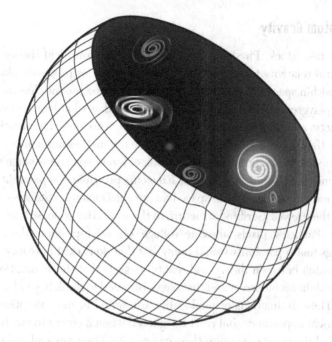

5.1. Boundary versus bulk. The universe may have a boundary, and if it does, physical processes occurring on the boundary can reproduce everything that is happening within the full volume of space. (Illustration by Jen Christensen)

equivalently by different numbers of spatial dimensions—then space can't be as fundamental as people used to think.

These peculiar implications are all the more astonishing when you consider their humble origins. Physicists took the simple observation that birds can safely perch on wires, extended it to gravity, and followed the logic where it led them. Gauge invariance, both in electromagnetism and gravitation, indicates a redundancy in our description of nature: multiple values of the electric potential are equivalent ways to describe a given situation; multiple values of the gravitational field correspond to the same relative arrangement of objects. Now we find that an entire dimension of space can be redundant. When space has a boundary, the entire universe can be collapsed onto that boundary, as though the full volume were superfluous. Space is unraveling before our eyes.

Quantum Gravity

Let's take stock. Physicists designed both quantum field theory and general relativity to explain the forces of nature as processes playing out within space, but the forces have proved to be too mischievous for that playground. The argument can be framed as what logicians call a reductio ad absurdum: make an assumption, work out its implications, show they're wrong, and conclude that the original assumption must be erroneous. When physicists try to think of forces spatially—when they assume that the universe satisfies Einstein's principle of separability, so that individual locations in space have an autonomous existence—they find themselves ascribing structure to the world that it doesn't actually have. Particles, pixels, absolute voltage, objective position: these are things that separability would imply, but that don't exist in reality. The mismatch between theory and reality suggests that the assumption of separability is incorrect and that physicists need to think beyond space.

These intimations of nonlocality are muted in most situations of practical importance, but can't be ignored when it comes to creating a unified theory—a quantum theory of gravity. Physicists realized early in the twentieth century that such a theory would almost certainly entail the dissolution of space. Crudely speaking, when you mash up a theory of atoms (quantum mechanics) with a theory of space (general relativity), you expect a theory of atoms of space. An "atom" in this context means the littlest possible chunk of space, which theorists initially visualized as a chessboard square. More than ordinary atoms of matter, these spatial cells fulfill the original meaning of the word "atom." They are truly indivisible. There's no such thing as smaller than such an atom, and position can't be defined within one, just as it doesn't matter exactly where in a chessboard square you place a chess piece; it might be located on square 9 or square 10, but never on square 9 3/4. Spatial atoms mark the logical limit of the reductionist program that has guided physics since ancient Greece.

The first proposals for such atoms had nothing to do with gravity per se but grew out of the millennia-old tension between the discrete and the continuum. As Zeno recognized long ago, a continuum has a lot—a *lot*—of room in it. If you tried to fill it with point-like particles, counting them out one by one, you could go on forever, adding an infinite number of particles, and still the continuum would be absolutely empty. Such a structure seems like overkill. It allows for more possible arrangements of matter than could ever be realized. "The continuum is more ample than the things to be described," Einstein wrote to a friend in 1916. This incongruity was bad enough in classical field theory and, as I mentioned in chapter 2, was one reason that physicists turned to quantum mechanics. Ironically, quantum field theory made it even worse. The theory predicted that fields vibrate spontaneously on all scales, filling up the continuum with an infinity of waves and causing forces to go completely out of control.

To restore order, many physicists proposed in the 1930s that fields don't vibrate below some threshold scale. Werner Heisenberg, in particular, had felt since the earliest days of quantum mechanics that the continuum must crumble on subatomic scales. He proposed a chessboard-like space that he called the *Gitterwelt*, or "lattice world." Yet this notion itself crumbled. The grid lines would privilege some directions over others, so that space would look different to a moving observer than to a stationary one, breaking the symmetry of relativity theory. In the late '40s, physicists reembraced the continuum. Infinite quantities, they reasoned, signaled merely that hitherto unnoticed fields come into play on small scales. By rounding out their selection of fields, they could create an internally consistent picture of electric, magnetic, and nuclear forces.

But gravity was another matter. It goes haywire on tiny scales, and in the early 1970s, theorists realized that no amount of rounding-out will wrestle it under control. So the continuum falls to pieces after all. The usual estimate for where this happens is known as the Planck scale, which is not just tiny but you've-got-to-be-kidding-me tiny—about as small in relation to an ordinary atom of matter as your body to the known universe. No microscope or particle accelerator built by human hands will ever see the Planckian filigree work. Nevertheless, it is implicit in things we do see, in the sense that empirically verified theories would collapse in a heap of inconsistency were it not for the breakdown of space at this level.

What exactly happens down there is still somewhat mysterious. Instead of a chessboard, space might be more like a watercolor painting whose brushstrokes bleed together, fuzzing out objects without creating the sharp divisions of a grid. Such a structure violates locality in the literal sense that you can't ever localize objects precisely. More important, this is an in for a penny, in for a pound situation. You can't change just one aspect of space—chop it into tiny morsels—without changing much else in order to keep everything behaving consistently. Would-be unifiers of physics picture a subatomic wonderland barely within the human capacity to comprehend.

In proposed quantum gravity theories, every weird thing you've ever heard about quantum particles becomes true of space. Just as particles can exist in more than one place at once, like Schrödinger's both-alive-and-dead cat, space can have multiple shapes at once. It was hard enough to demarcate position when space could change shape. Now space has no shape at all, just a blur of possibility, and position becomes undefinable. The web of cause and effect threatens to disintegrate. String theorists, who infer the properties of submicroscopic space from the behavior of the loops of energy that inhabit it, think that notions such as size and dimension become fluid. As you shrink a string down, it reaches some minimum size and then starts to *grow* again—like a balloon that, when squeezed in one place, inflates somewhere else, except that the somewhere else is an entirely new dimension of space. "Ordinary notions of spacetime are simply no longer valid in string theory," Joe Polchinski says.

In short, locality fails in every sense of the word. Still, these shenanigans are occurring way down on the Planck scale, or close to it. When you take the big-picture view, space should be just fine. Trillions upon trillions of spatial atoms blur together into a single unbroken expanse. Those atoms are mutually independent, so locality should hold; each part of space should have an independent existence. Or at least that has been the conventional wisdom.

Theorists who came of age in the 1990s, such as Fotini Markopoulou, found this line of reasoning unpersuasive. Could the grandest enterprise in all of theoretical science—unifying physics—really have such piddling consequences for the world at large? She and others got to thinking that our traditional notions of space dissolve not just deep inside particles, but also in the macroverse, across millions of kilometers of open space and perhaps even on the scale of the observable universe. "If quantum gravity is as fundamental as we think it is, if it's something that's about the very structure of spacetime, it's just not obvious to me that the signature of that thing has to be something very small," Markopoulou says.

She has a point. Atoms of matter are pretty small, too, but we are always being confronted with their existence. For instance, materials made of atoms can metamorphose: graphite can turn to diamond, not to mention combine with other species to create chemically distinct substances. By analogy, if space consists of atoms, it stands to reason that those atoms might rearrange themselves into something other than space. Such a transformation would be extremely unpiddling.

Although physicists often describe quantum mechanics as a theory of the microworld, they admit this is a white lie. That may be how their predecessors originally conceived of the theory, but it doesn't reflect their present understanding. As far as people nowadays can tell, quantum mechanics is a theory of the *world*, period. Experimentalists have yet to see any range of sizes where nature ceases to behave quantum-mechanically. The human body is every bit as quantum as an electron. You don't notice people around you pulling quantum stunts such as existing and not existing at the same time, but that's not because they're big, per se. It's because the human body is an open system that vigorously interacts with its environment, dispersing the distinctively quantum effects like dandelion fluff in the breeze. Under the right conditions, you can see such effects with the unaided eye. (Experimenters can't yet manage this for a whole human but have witnessed, for example, miniature tuning forks both vibrating and not vibrating.) And the same should be true for spacetime.

Back to Black

Because black holes are roiled both by quantum effects and by intense gravity, they're the poster children for quantum gravity. And we're not talking piddling atoms anymore. These are monsters that could swallow a solar system. The most obvious place where quantum gravitational effects operate is at the center of these objects, where matter becomes so tightly packed that it pushes up against the density limit imposed by the atomized structure of space. But the perimeter of the hole presents its own problems, such as the paradoxical behavior discovered by Hawking, which is a red flag that locality is breaking down on large scales as well as small. "It's been the story in gravity that the singularity is really something to worry about," Giddings says. "But now, with the information paradox, something funny has to happen all the way out at the horizon. Focus on the singularity and short-distance issues has been a little misplaced. There are long-distance issues."

The various types of nonlocality I've been talking about converge in black holes. For instance, Hawking's paradox has connections to quantum entanglement that physicists have only lately begun to explore. General relativity says a black hole is not a material object like a star or planet. It's mostly vacuum—a pit of nothingness. An unfortunate soul falling into one sees only empty space. Yet the bleak façade conceals a frenzy of activity. Quantum field theory says a vacuum may be devoid of particles, but not of *everything*; the fields are still there, merely becalmed. Superentanglement among all the regions of space ensures that vibrations in the fields negate one another precisely, leaving only an uneasy silence, like the quiet you hear with noise-canceling headphones—not so much the absence of sound as the active repression of it.

If anything throws off this balance, the latent activity of the vacuum erupts into the open. And indeed that's what happens for the wise person who resists falling into the hole. Relative to the person falling in, the outside observer must keep accelerating outward—for instance, by throttling up his rocket engines to fight the hole's gravitational suction. By the logic of relativity theory, the rocket man experiences the passage of time differently than the infalling victim does, and he will also experience the field's vibrations differently, because vibration is a process that occurs within time. He does not, in fact, think they cancel out. The fields aren't becalmed; particles are flying off in every direction. Some escape into deep space, draining the hole of its energy until it winks out of existence. That observer has taken off the noise-canceling headphones and heard the roar that lies on the other side of silence.

This isn't the paradox yet, just a new incarnation of Hawking's argument that black holes decay. The real question is: What becomes of objects that fell into the black hole? The hole's demise should liberate them, but how? Perhaps they make their getaway under cover of those outgoing particles. If you collected the particles and pieced them to-gether, you might be able to reconstitute all the objects that the hole devoured. For that to work, the particles would have to be entangled with one another, so that collectively they retain all the qualities of whatever fell in. And there's the rub. We've insisted that entanglement do double duty: preserve the emptiness of the vacuum as perceived by the infalling observer while keeping alive the memory of the hole's victims. That may be asking too much of it. Entanglement is a limited resource in much the same way energy is, and there doesn't seem to be enough to perform both tasks. Either the hole is not a vacuum after all (in which case general relativity theory is badly wrong) or falling into the hole is irreversible (in which case quantum field theory is toast). Theorists have sought ways to wriggle out of this dilemma; for instance, perhaps entanglement can be structured in a way that balances both needs. But if not, the impasse demands new physics—some nonlocal effect that makes the hole a two-way street or erases the distinction between inside and outside. "It means a violent breakdown of locality," Polchinski says.

Another conundrum arises when you ask how the hole stores in-falling objects (or the pieces into which it rips them) until their even-tual release. The hole must grow to accommodate those objects, and it doesn't grow by the amount locality would lead you to expect. Black holes should be like suitcases. Imagine you have an old suitcase with 10 compartments, each holding one pair of socks, and then you buy a new bag with twice the linear dimensions, for eight times the volume. You'd expect the new bag to have 80 compartments and hold 80 pairs of socks. What locality means in this context is that individual compart-ments have the same size, and the new bag just gives you more of them. The advantage isn't just that you can pack more clothes, but that you can pack a greater variety of clothes. Physicists say that the larger bag has a greater entropy. Entropy is commonly described as disorder, but it can also be thought of as the potential for variety—in this case, the multiplicity of ways you could pack the bag without causing it to weigh more, bulge out, or otherwise change its external appearance.

By this reasoning, doubling the radius of a black hole should increase its volume eightfold, and its storage capacity should scale up propor-tionately. But that's not what happens. When a black hole doubles in radius, its mass only doubles, while its entropy increases fourfold. The hole has less storage capacity than theories based on locality predict. It's like opening your big new suitcase to find only forty compartments, each holding a single sock rather than a pair, for a total of twenty pairs rather than the eighty they promised you at the shop. You'd feel cheated.

The fourfold increase in entropy means that the internal complexity of the hole is growing not with its volume, as locality would predict, but with the area of its horizon. In effect, increasing the width and length of the hole adds to its storage capacity, but extra height has no effect, as though the height dimension were illusory. The thing looks 3-D, but acts as though it were 2-D. And what is true for black holes is true for everything else, because anything can be turned into a black hole by squeezing hard enough. Space is pulling a kind of bait and switch on us. It offers so much room to hold stuff—indeed, isn't holding stuff supposed to be what space is all about? But when you try to fill it up, you find yourself strangely unable to do so. You begin to wonder whether the space is really there or something is tricking you, like a wall of mirrors that makes a cramped Manhattan studio look like a penthouse suite.

•

The chicanery of black holes is a smaller-scale version of what happens in universes with a boundary. Again, a volume of space can be collapsed onto its exterior surface. This insight doesn't just take down our conception of space, but tells us what might replace it.

Technically, the boundary and volume are equivalent, neither having precedence over the other. But many physicists believe the two aren't truly identical—that the boundary is the fundamental reality and the volume is derived from it. This proposition is known as the "holographic principle" because the boundary is acting like a flat sheet of film able to evoke a 3-D scene, analogous to that little silvery image on the back of your credit card. A hologram works because it's an elaborately patterned photograph that captures the depth cues in a scene. Likewise, fields on the boundary of space are so elaborately patterned that they can reproduce the entire universe.

If they were giving out prizes for the most confusing physics metaphor ever, the holographic principle would be a strong contender. To be fair, it captures the way that a dimension of space seems to go poof. Also, holograms have a mystique. They make you go "Whoa!" and that's exactly the reaction physicists want you to have when dimensions of space go missing. On the downside, what the heck does the holographic principle actually mean? How could the universe around us be a holographic projection? What's the projector? The key, I think, is not to get caught up in the minutiae of the metaphor. The point is just that the space we observe could be a product of some underlying structure. When we walk across a room, we are not gliding passively through a preexisting expanse. Something is *happening*. There is machinery at work, a grinding of gears deep within nature, to produce the experience of being "here" and being "there." When you stretch an arm to grasp a pencil and it's just outside your reach, something is acting to thwart you, creating what you perceive as distance. And when we ask what that machinery could be, we have arrived at the outermost frontier of modern physics. "At first sight, it's pretty obscure how that happens," says Daniel Kabat, a theorist at Lehman College in New York. "Even at second sight, it's still very obscure . . . Physics is local. It means something to say we're in this room talking, or objects are at different but well-defined positions. Understanding how that emerges from this framework is pretty obscure."

Physicists, like the rest of us, grasp abstract concepts by applying them to concrete examples, and the holographic principle rose above the level of curiosity only when Juan Maldacena fleshed out his idea of AdS/CFT duality. "AdS" stands for "anti-de Sitter," meaning the interior of a higher-dimensional ball, a simple type of bounded universe. "CFT" stands for "conformal field theory," meaning the surface of the ball. "Duality" means that these two realms are equivalent. By making measurements on the boundary, you can work out what's going on in the bulk.

Consider what this means. A realm governed by gravity (described by Einstein's general relativity and its quantum elaboration) is equivalent to a realm governed only by nongravitational forces (described by quantum field theory with gauge invariance). Maldacena's analysis therefore achieves the long-sought unification of these two branches of physics, at least for this idealized ball-shaped universe. Historically, physicists have sought unity by zooming in to look for smaller-scale building blocks such as atoms and particles. In Maldacena's approach, they don't look for things that are smaller, but things that don't even exist in the same space.

I have met few theorists who aren't in awe of Maldacena and what he accomplished (although they may disagree on whether the theoretical structure has anything to do the real world). One tells me he once gave Maldacena a lift in his car and was utterly terrified. If he crashed, he worried, nothing else he had done in his life would ever be remembered. He would go down in history as the man who killed Juan Maldacena.

Using the holographic principle, theorists can investigate what kinds of processes might give rise to space that wasn't already there. Notwithstanding the usual prudent scientific disclaimers—more research is needed and so forth—a solution to the mysteries of nonlocality is beginning to present itself. Through the haze, we can see the outlines of the mountain.

6

Spacetime Is Doomed

When the philosopher Jenann Ismael was ten years old, her father, an Iraqi-born professor at the University of Calgary, bought a big wooden cabinet at an auction. Rummaging through it, she came across an old kaleidoscope, and she was entranced. For hours she experimented with it and figured out how it worked. "I didn't tell my sister when I found it, because I was scared she'd want it," she recalls. As you peer into a kaleidoscope and turn the tube, multicolored shapes begin to blossom, spin, and merge, shifting unpredictably in seeming defiance of rational explanation, almost as if they were exerting spooky action at a distance on one another. But the more you marvel at them, the more regularity you notice in their motion. Shapes on opposite sides of your visual field change in unison, and their symmetry clues you in to what's really going on: those shapes aren't physical objects, but images of objects—of shards of glass that are jiggling around inside a mirrored tube. "There's a single bead of glass that's being redundantly represented in different parts of the space," Ismael says. "If you focus in on the larger embedding space, the physical description of the three-dimensional kaleidoscope, you've got a straightforward causal story. There's a piece of glass, the piece of glass is being reflected along the mirrors, and so on." Seen for what it really is, the kaleidoscope is no longer mysterious, though still pretty awesome.

Decades later, while preparing a talk on quantum physics, Ismael thought back to the kaleidoscope and went out to buy a fancy new one, a shiny copper tube in a velvet case. It was, she realized, a metaphor for nonlocality in physics. Maybe particles in an entanglement experiment or galaxies on the farthest reaches of known space act strangely because they're really projections—or, in some other way, secondary creations—of objects existing in a very different realm. "In the kaleidoscope case, we know what we have to do: we have to see the whole system; we have to see how the image space is created," Ismael says. "How do we construct an analogue of that for quantum effects? That means seeing space as we know it—everyday space in which we view measurement events located at different parts of space—as an emergent structure. Maybe when we're looking at two parts, we're seeing the *same* event. We're interacting with the same bit of reality from different parts of space."

She and others question the assumption, made by nearly every physicist and philosopher from Democritus onward, that space is the deepest level of physical reality. Just as the script of a play describes what actors do on stage, but presupposes the stage, the laws of physics have traditionally taken the existence of space as a given. Today we know that the universe has more to it than things situated within space. Nonlocal phenomena leap out of space; they have no place in its confines. They hint at a level of reality deeper than space, where the concept of distance ceases to apply, where things that appear to lie far apart are actually nearby or perhaps are the same thing manifested in more than one place, like multiple images of a single shard of kaleidoscopic glass. When we think in terms of such a level, the connections between subatomic particles across a lab bench, between the inside and the outside of a black hole, and between opposite sides of the universe don't seem so spooky anymore. Michael Heller, a physicist, philosopher, and priest at the Pontifical Academy of Theology in Krakow, Poland, says: "If you agree that the fundamental level of physics is not local, everything is natural, because these two particles which are far apart from each other explore the same fundamental nonlocal level. For them, time and space don't matter." Only when you try to visualize these phenomena in terms of space—which is forgivable, because it's hard for us to think in any other way—do they defy comprehension.

The idea of a deeper level seems natural because, after all, it is what physicists have always sought. Whenever they can't fathom some aspect of our world, they assume they must not yet have gotten to the bottom of it all. They zoom in and look for the building blocks. How mysterious it is, for example, that liquid water can boil to steam or freeze to ice. Yet these transformations make perfect sense if liquid, vapor, and solid are not elemental substances, but distinct forms of a single fundamental substance. Aristotle took the states of water to be diverse incarnations of so-called prime matter, and the atomists— presciently—thought they were rearrangements of atoms into tighter or looser structures. En masse, the building blocks of matter acquire properties that, individually, they lack. A molecule of water is not wet, and an atom of carbon is not alive, but lots of them, coming together in the right way, can be. Likewise, space might be built of pieces that are not themselves spatial. Those pieces might also be disassembled and reassembled into nonspatial structures such as the ones that black holes and the big bang are hinting at. "Spacetime can't be fundamental," says the theorist Nima Arkani-Hamed. "It has to come out of something more basic."

This thinking completely inverts physics. Nonlocality is no longer the mystery; it's the way things really are, and *locality* becomes the puzzle. When we can no longer take space for granted, we have to explain what it is and how it arises, either on its own or in union with time. Clearly, constructing space isn't going to be as straightforward as melding molecules into a fluid. What could its building blocks possibly be? Normally we assume that building blocks must be smaller than the things you build out of them. A friend of mine and his daughter once erected a detailed model of the Eiffel Tower out of popsicle sticks; they hardly needed to explain that the sticks were smaller than the tower. When it comes to space, though, there can be no "smaller," because size itself is a spatial concept. The building blocks cannot presume space if they are to explain it. They must have neither size nor location; they are everywhere, spanning the entire universe, and nowhere, impossible to point to. What would it mean for things not to have positions? Where would they be? "When we talk about emergent spacetime, it must come out of some framework that is very far from what we're familiar with," Arkani-Hamed says.

Within Western philosophy, the realm beyond space has traditionally been considered a realm beyond physics—the plane of God's

existence in Christian theology. In the early eighteenth century, Gottfried Leibniz's "monads"—which he imagined to be the primitive elements of the universe—existed, like God, outside space and time. His
theory was a step toward emergent spacetime, but it was still metaphysical, with only a vague connection to the world of concrete things.
If physicists are to succeed in explaining space as emergent, they must
claim the concept of spacelessness as their own.

Einstein foresaw these difficulties. "Perhaps . . . we must also give
up, by principle, the space-time continuum," he wrote. "It is not unimaginable that human ingenuity will some day find methods which
will make it possible to proceed along such a path. At the present time,
however, such a program looks like an attempt to breathe in empty
space." John Wheeler, the renowned gravity theorist, speculated that
spacetime is built out of "pregeometry," but admitted this was nothing
more than "an idea for an idea." Even someone as irrepressible as Arkani-
Hamed has had his doubts: "These problems are very hard. They're outside our usual language for talking about them."

After decades of effort, though, he and others have risen to Einstein's challenge. In this chapter, I will synthesize ideas from a wide
variety of research programs into a new picture of what space is and
what might underlie it. "For 2,000-plus years, people asked about the
deep nature of space and time, but they were premature," Arkani-
Hamed has said. We've finally arrived at the epoch where you *can* pose
the questions and hope for some meaningful answer."

The Glue of Reality

Rather than take space for granted, let's rethink what we really mean
by it and how we might do without it. According to the classical laws of
physics developed by Newton and taught today in high school, the
positions and velocities of objects characterize the world fully. Those
quantities at one instant reflect how everything got to where it is and
determine what it'll do in the future. You can think of the world as one
of those giant Rube Goldberg contraptions that people build in an
abandoned warehouse and post to YouTube: once everything is in

place, you set it in motion and all else follows. If you arrange atoms in the shape of a frog and flick them with the right velocity, the frog will live; it needs no additional galvanic spark.

Classical physics says things will slow down, speed up, or veer off their original courses because they're exerting forces on one another, such as electrical and gravitational forces. The force strength depends on the things' relative positions—in particular, how far apart they are. It makes no difference what may or may not have gone on between those things in the past. They may once have rubbed up against each other, but once they parted company, the attraction or repulsion between them was no stronger and no weaker than that of anything else the same distance apart. Space is brutally egalitarian. When you become separated from your lover, the two of you retain no tighter a physical connection than do two lumps of coal.

In this way, space serves as the organizing principle of the natural world—the glue that binds the universe together, as the English physicist Julian Barbour has put it. Physical objects do not interact willy-nilly; their behavior is dictated by how they are related to one another, which depends on where they lie in space at a given time. This structuring role is easiest to see in the classical laws of mechanical motion, but also occurs in field theories. The value and rate of change of a field at different points in space fully determine what the field does, and points in the field interact only with their immediate neighbors.

I don't want to leave the impression that theories formulated within space are always able to capture the world, when clearly they don't. I've mentioned some of their limitations in previous chapters. For instance, inconsistencies arise when discrete objects and continuous fields intermix, and both Newton's and Einstein's laws of motion go kaput at so-called singularities, where physical quantities such as speed or density become infinite. Despite these exceptions, though, physicists have always situated the elements of their theories within space. Even in quantum mechanics, if you put aside the mysteries of entanglement, the interactions within a system are limited by the spatial arrangement.

But we can invert this reasoning. Physicists and philosophers can *define* space as the fact that the natural world has a very specific structure to it. Instead of saying that space brings order to the world, you can say that the world is ordered and space is a convenient notion for describing that order. We perceive that things affect one another in a certain way and, from that, we assign them locations in space. This structure has two important aspects. First, the influences that act on us are hierarchical. Some things affect us more than other things do, and from this variation we infer their distance. A weak effect means far apart; a strong effect implies proximity. The philosopher David Albert calls this definition of distance "interactive distance." "What it means that the lion is close to me is that it might hurt me," he says. This is the opposite of our usual mode of thinking. Rather than cry, "Watch out, the lion is close, it might pounce!" we exclaim, "Uh-oh, the lion might pounce on me; I guess it must be close."

The second aspect of the spatial structure is that diverse influences are mutually consistent. If a rhinoceros is also able to hurt me, it must be close, too. And if both a lion and a rhino are able to hurt me, then the lion and rhino should also be able to hurt each other. (Indeed, my survival depends on it.) From this patterning of influences, we extract space. If the threat posed by predators couldn't be expressed in terms of spatial distance, space would cease to be meaningful. A less morbid example is triangulation. The signal bars on your mobile phone indicate the strength of the phone's connection to a cell tower and therefore your distance from that tower. In an emergency, the phone company can locate your phone by measuring your signal at several towers and using triangulation or the related technique of trilateration. The fact that the measurements converge on a single location is what it means for you to *have* a location.

•

The nice thing about defining space in terms of structure is that it sidesteps some of the long-running disputes over the nature of space. The ancient atomists conceived of space as a thing in its own right, whereas Aristotle deemed it an abstraction that describes how the contents of the universe are packed together. Either way, though, space reflects a structure that the natural world possesses. If the atomists were right and space has an independent existence, it must be highly ordered, like a neatly woven fabric, so that it can serve the functions that physics demands of it. If space is merely an abstraction, then the contents of the universe must fit together in just the right way to give meaning to the abstraction.

On the face of it, we haven't gained anything by giving primacy to order; we've merely traded one mystery for another, because now we've got to explain the order. The theoretical physicist Lee Smolin of the Perimeter Institute calls this the "inverse problem." But one person's problem is another's opportunity, because now we can imagine what the universe would be like if it *weren't* ordered in the requisite way. Then it might not be spatial anymore. Instead of thinking of space as an absolute necessity, we can regard it as one of the possible states of the universe, just as ice is one of the possible states of water. The ice analogy isn't bad, actually. Water is solid over a narrower range of conditions than it is gaseous. Likewise, space may be the exception, not the rule; most proposed unified theories of physics suggest that the vast majority of the universe's possible states are nonspatial. "Where spacetime exists is very nongeneric," says Moshe Rozali, a string theorist at the University of British Columbia. "It requires some special conditions." In the twilight between order and disorder, space and spacelessness, perhaps we can find a comprehensible explanation for the nonlocal phenomena that physicists have puzzled over.

Much the same applies to time, too. I've been focusing on space because most researchers do so; the lessons of nonlocality for time are much less clear. Smolin and the philosopher Tim Maudlin have recently written books to argue that time isn't emergent, a prospect I will return to later. Whatever its foundational status, time plays a powerful organizing role in the universe, and as with space, this structure has two aspects. First, it is hierarchical. Events can be closely related, distantly related, or unrelated. They happen in a logical sequence, one event leading to the next, according to the laws of physics. For example, in a classic psychological test, you're shown a series of pictures and have to arrange them to tell a story. The dog shakes itself dry; the dog runs into a puddle; the dog slips its leash. Out of order, these scenes make for a surreal story; sorted correctly, they become an episode that any dog owner will recognize. Even a nonlinear narrative such as *Slaughterhouse-Five* or the film version of *Cloud Atlas* has a chronological logic.

In one of the last essays he wrote before his death, Leibniz turned this thinking upside down. Time isn't the reason that events are structured, he argued. It's the *consequence* of the fact that they're structured. You can derive time from the sequence of cause and effect. One event that leads to another must precede it in time, and the number of intermediate steps determines how much time passes between them. Einstein's theory of relativity suggests that Leibniz was onto something, because it gives cause-effect relations a central function in physics. One event can trigger another only if light has enough time to travel between them. Reversing the logic, if one event can trigger another, the two must occur within such-and-such a distance of each other. In fact, you could map out the entire world from its web of cause-effect relations—you could put everything in the right place just by knowing what has to give rise to what.

A number of physicists have developed this idea into a model of spacetime as a "causal set." Technically, the web of relations does leave some ambiguity: its tells you that one event occurs earlier than another, but not *how much* earlier; it provides no scale. But causal-set theorists argue that space and time already have a natural scale if they are built up out of discrete units—"atoms" of space—as I talked about in the previous chapter. Distance is then determined by counting the number of such atoms.

For our purposes, what's important is that the web of cause-effect relations must be highly ordered if it is to re-create space and time. "Almost all the causal sets look nothing like spacetime," says Fay Dowker, a causal-set pioneer. By analogy, the makers of that psychological test can't just throw any scenes at you and expect you to make sense of them; they must be scenes that dovetail together. When the scenes do have an order, time is a label for where each one comes in the sequence.

In addition to creating a sequence or hierarchy of events, the temporal structure ensures the mutual consistency of different processes. To be more specific, forget that you ever heard of time, and think about clocks. Looking around, you see these very useful things called clocks. They come in handy for baking and exercise regimens. If you synchronize the clocks in your house, you can count on them to stay in sync, at least approximately. Because all the clocks display the same numbers, you call those numbers "the time." We seldom stop to think what a remarkable fact this regularity is. Why should a kitchen timer tell you when bread will rise? Why should Big Ben spin in sync with the Earth? These are utterly different systems: electronic oscillator, fermenting yeast; swinging pendulum, giant rotating ball. Yet they march to the same drummer.

Julian Barbour argues that these systems track one another's motion because they are all interlocked, as if they were gears in a larger clockwork. By this reasoning, time does not precede the universe, enforcing its internal consistency, but comes from within. Ernst Mach, who was a proponent of Leibniz's views, wrote: "All things in the world are connected with one another and depend on one another . . . Time is an abstraction, at which we arrive by means of the changes of things; made because we are not restricted to any one *definite* measure, all being interconnected . . . We reach our ideas of time in and through the interdependence of things on one another. In these ideas the profoundest and most universal connection of things is expressed." Time, like space, draws its meaning from a special type of harmony in nature.

When Space Is Not the Place

You don't need a complex system of moving, interacting parts to appreciate the organizing power of space. Consider the geography of a country. You can think of the cities as laid out on a map or you can express their mutual spatial relations with a mileage chart, like those found in paper road maps and atlases—one of those rectangular or triangular grids giving the distances between pairs of cities. What's interesting is that the chart contains hidden patterns within it, just as jigsaw puzzle pieces look unrelated when you dump them out of the box but show their affinity as you fit them together.

Suppose you have twenty cities. The chart contains four hundred numbers. To fill in the data, a map company such as AAA hires drivers to travel from city to city and jot down the odometer mileage or take GPS readings. In terms of real information content, the chart is highly

Intercity mileage and travel time chart (miles on top line, travel time hh:mm on bottom line):

To \ From	Atlanta	Boston	Chicago	Cleveland	Dallas	Denver	Detroit	Las Vegas	Los Angeles	Miami	Minneapolis	New Orleans	New York City	Orlando	Phoenix	St. Louis	Salt Lake City	San Francisco	Seattle
Boston	1106 / 16:18																		
Chicago	719 / 10:34	987 / 14:24																	
Cleveland	709 / 10:21	639 / 9:14	346 / 5:11																
Dallas	783 / 11:27	1772 / 25:40	967 / 13:56	1194 / 17:17															
Denver	1404 / 20:14	1975 / 28:28	1005 / 14:28	1334 / 19:19	879 / 12:50														
Detroit	721 / 10:26	814 / 11:53	282 / 4:17	176 / 2:47	1194 / 17:09	1270 / 18:11													
Las Vegas	2040 / 29:12	2718 / 39:11	1748 / 25:11	2077 / 30:02	1316 / 18:56	747 / 10:57	2012 / 28:54												
Los Angeles	2218 / 31:43	2988 / 42:55	2016 / 28:55	2345 / 33:46	1437 / 20:28	1016 / 14:41	2281 / 32:36	270 / 3:52											
Miami	664 / 9:47	1505 / 22:08	1391 / 20:26	1243 / 18:09	1363 / 19:43	2073 / 30:04	1390 / 20:16	2676 / 38:31	2741 / 39:12										
Minneapolis	1128 / 16:10	1402 / 20:24	408 / 5:57	761 / 11:14	988 / 14:15	915 / 13:06	697 / 10:07	1657 / 23:49	1926 / 27:33	1797 / 26:00									
New Orleans	468 / 6:55	1526 / 22:19	926 / 13:19	1065 / 15:33	520 / 7:43	1396 / 20:23	1064 / 15:26	1833 / 26:31	1898 / 27:11	866 / 12:32	1295 / 18:37								
New York City	883 / 12:57	218 / 3:40	790 / 11:41	463 / 6:53	1549 / 22:21	1779 / 25:45	614 / 9:06	2521 / 36:28	2790 / 40:12	1290 / 18:49	1206 / 17:40	1303 / 18:56							
Orlando	438 / 6:26	1298 / 19:06	1163 / 17:07	1035 / 15:09	1136 / 16:26	1846 / 26:47	1163 / 16:59	2449 / 35:15	2514 / 35:56	240 / 3:51	1570 / 22:43	639 / 9:15	1083 / 15:50						
Phoenix	1846 / 26:32	2700 / 38:45	1803 / 25:56	2067 / 29:42	1065 / 15:19	909 / 12:59	2036 / 29:31	291 / 4:30	372 / 5:17	2369 / 34:02	1791 / 25:33	1527 / 22:01	2466 / 35:43	2142 / 30:46					
St. Louis	554 / 8:02	1193 / 17:13	297 / 4:21	560 / 8:12	629 / 9:25	850 / 12:17	530 / 7:59	1596 / 22:59	1825 / 26:03	1223 / 17:52	535 / 8:21	675 / 9:51	959 / 14:15	996 / 14:35	1502 / 21:34				
Salt Lake City	1921 / 27:28	2369 / 33:55	1399 / 20:00	1728 / 24:43	1404 / 20:10	524 / 7:54	1664 / 23:42	419 / 6:03	688 / 9:46	2590 / 37:19	1309 / 18:38	1921 / 27:44	2173 / 31:13	2359 / 34:04	651 / 9:56	1366 / 19:31			
San Francisco	2488 / 36:00	3103 / 44:26	2133 / 30:26	2462 / 35:17	1815 / 25:58	1268 / 18:03	2397 / 34:09	570 / 8:27	380 / 5:33	3119 / 44:41	2042 / 29:04	2277 / 32:40	2906 / 41:43	2891 / 41:21	750 / 10:45	2099 / 29:56	735 / 10:35		
Seattle	2696 / 38:29	3057 / 43:54	2063 / 29:31	2417 / 34:48	2199 / 31:29	1331 / 19:00	2352 / 33:40	1257 / 17:54	1142 / 16:29	3365 / 48:20	1657 / 23:42	2717 / 39:02	2862 / 41:14	3138 / 45:02	1512 / 21:40	2142 / 30:32	840 / 12:01	815 / 11:58	
Washington, DC	638 / 9:18	443 / 7:00	697 / 10:35	367 / 5:53	1330 / 19:13	1680 / 24:31	520 / 7:57	2427 / 35:14	2675 / 38:14	1056 / 15:12	1112 / 16:35	1084 / 15:54	228 / 3:41	849 / 12:12	2352 / 33:42	839 / 12:27	2080 / 30:04	2813 / 40:37	2768 / 40:08

6.1. Intercity mileage and travel time chart. (© AAA, used by permission)

redundant. The distances obey very specific rules that mathematicians call "distance axioms." To begin with, the twenty numbers running along the diagonal are zero—the distance from each city to itself. Of the remaining numbers, half are repeats, since distances are symmetrical: a car driving from Dallas to Salt Lake City covers basically the same ground as a car going the other way. Indeed, most map companies leave off this redundant information and show only the remaining triangle of numbers.

Even the 190 quantities in that triangle are not fully independent of one another, because you can boil them down to 60 values representing each city's coordinates—latitude, longitude, and altitude—as well as an additional value for Earth's radius when cities are far enough apart that the planet's curvature becomes a factor. Finally, you can lose several more numbers because the conventions used for the co-ordinate system (such as taking the prime meridian to be 0 degrees longitude) don't matter for the purposes of driving distances. That knocks out another six quantities. So now you're down to fifty-five. The four hundred numbers you started with are just various arithmetic combinations of fifty-five numbers. That may not be obvious from looking at the chart, but you know it has to be true because you could reverse the process. You could start with the cities' coordinates, mark their locations on a map, and use trigonometry to calculate the intercity distances.

So the chart is highly ordered. This is what it means for the cities to be situated within space. Spatial coordinates are a highly economical way of capturing the possible mutual relations among things. In the above example, we had twenty cities and four hundred intercity distances that reduce to fifty-five unique numbers. The more things you have, the more impressive the savings are. For one hundred cities, the chart con-tains ten thousand intercity distances, reducing to 295 numbers. For all the world's cities, or all its towns, or all its geographical features of any type, the raw distance data would consume a hard drive even though the positions of those features can be concisely expressed on a single map. "That's what space is," Barbour says. "It's data compression on a massive scale."

The reason the compression is so powerful is locality. Locality means that the whole is the sum of its spatial parts, and in this context, that means every journey is a series of smaller steps. You can build up long distances from shorter intermediate ones, so you don't need to specify each and every pair of directions. For instance, the chart might tell you it's 900 miles from Dallas to Denver and 500 miles from Den-ver to Salt Lake City, so you don't need to be told it's at most 1,400 miles from Dallas to Salt Lake.

Suppose this weren't the case—suppose the data in the chart weren't so highly ordered. If I fill in a chart with four hundred random numbers and ask you to mark their locations on a map, you'll almost certainly fail. For instance, the chart might tell you it's nine hundred

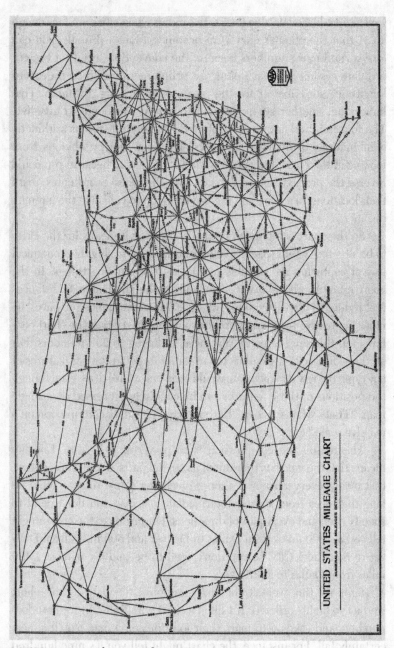

6.2. U.S. intercity mileage map from 1939. (© AAA, used by permission)

miles from Dallas to Denver, five hundred miles from Denver to Salt Lake City, and eight thousand miles from Dallas to Salt Lake. Now, that doesn't make much sense, does it? These data put Salt Lake City in two different places, depending on whether you drive straight from Dallas or stop off in Denver. The situation is like an April Fool's joke in which your friend mixes together pieces from different jigsaw puzzles and gives them to you to assemble. You'll struggle to fit the pieces together until it dawns on you that your supposed friend is a cruel prankster.

Under such circumstances, position becomes meaningless. *Space* becomes meaningless. It's not a useful way to describe the relations among places anymore. But that doesn't mean the relative arrangement of cities is incomprehensible. Even if you can't place the cities on a map, you can fall back on the full mileage chart. In other words, you can use what philosophers call the "unmediated" distances, the ones that directly link pairs of cities and can't be reduced to a series of shorter hops. This isn't an entirely hypothetical situation. When I first drove in Boston, I had to learn to distrust my spatial awareness, because it kept getting me lost. In that maze of one-way streets and amoeba-shaped "squares," you routinely have to go west to go east, or get in the left lane to turn right. It does you no good to know where places are located from a bird's-eye view; instead, you need to robotically follow directions for where and how to turn. To a driver, Boston is a nonspatial city.

•

It's really not so strange that networks of relations can burst out of space. This is the case, after all, for human relationships. Our social lives are too tangled to be laid out on a spatial map. Not that people don't try. Family trees translate genetic and conjugal closeness into spatial closeness, and online social networks have spawned similar attempts. The Wolfram Alpha website used to be able to map out your Facebook friends network, using dots to symbolize your friends and lines to connect those who have friended each other. Spatial distance on the map represents social proximity as judged by the number of friends people have in common. Typically, your friends cluster into distinct social circles: family, classmates, workmates, ultimate Frisbee teammates,

6.3. Facebook friends network. (Wolfram Alpha LLC, 2014, used by permission)

fellow Radiohead groupies, and so on. If these people go to the same party, they might congregate in different corners of the room, and the figurative distance between them translates into literal distance.

When I first generated a Facebook graph, I noticed a stray link between my physics colleagues and my music friends, revealing that a theorist I'd worked with shared my passion for Cuban dance music. Finding unexpected connections is half the fun of these graphs, but does expose the limitations of the spatial metaphor. There's no consistent way to assign that theorist a location on my graph. Like Salt Lake City in the above example, he occupies two different places, corresponding to two social circles. And the failure deepens when you consider everything these graphs leave off. Two Facebook "friends" may have never met or spoken, yet the diagram links them as if they were BFFs. One person might have an unrequited crush on the other, and still a line connects them. To capture these other dimensions of human relationships, you can festoon family trees with symbols: thick line for a close bond, zigzags for hostility, and so on. Such diagrams, known as genograms, are popular among psychologists, social workers, and people struggling to follow *Game of Thrones*. The symbols compensate for the failure of the spatial metaphor.

In some cases, people organize themselves so that their social network becomes radically streamlined, and these situations let us see how space might emerge from spacelessness. A structure can form where none existed before. That can happen in two ways: build up or cut back. People might start as atomized individuals who begin interacting, like your grandmother who finally got on Facebook and signs up all her friends. Or they might start with a mess of existing relations and prune them, like a social butterfly who friended everyone he met, realizes he doesn't know who half of them are, and does a friends purge.

The army, for example, restricts socializing across ranks, on the assumption that familiarity might breed contempt. Consequently, difference in rank is analogous to spatial separation: a private is distant from a colonel in much the same way that Dallas is distant from Salt Lake City. Information flows up and down the chain of command just as a person driving from Dallas to Salt Lake City must pass through intermediate points. Because of this structure, a military hierarchical chart is a fair representation of social relations in the military.

The army's structure is imposed by military discipline, but in other cases the order develops spontaneously from within. A classic example is the market economy. We routinely speak of "the economy" as though it were a conscious being rather than millions of people making rash decisions with their money. And in a sense it is, because collective arrangements transcend the people who create them. In isolation, individuals don't put price tags on goods, because they have no one to buy from or sell to. Price becomes important when people come together and trade. Depending on their haggling skills, the price varies from person to person and place to place. Some plucky entrepreneur takes advantage of these variations to buy low and sell high; in so doing, that person helps to even out the supply and therefore harmonize the prices.

This kind of self-organizing happens all the time in physics. For instance, a single water molecule has no temperature. Temperature becomes meaningful when molecules collide and exchange energy. If you mix cold and hot water, the cold warms up, the hot cools off, until they equalize. Before equilibrium, the water is characterized by two temperatures; afterward, by a single value. From complexity comes simplicity. The complexity remains latent, though. You can tell it's there whenever the temperature fluctuates or water undergoes a transformation such as boiling in a teakettle. Physicists commonly use these deviations from standard behavior as windows into the microscopic composition of materials.

The same might go for space, too. The basic building blocks of nature might be capable of a tangle of relationships that would fill a celebrity gossip rag. Through some organizing mechanism or simply the play of averages, those relationships become regimented, so that they can be laid out on a spatial grid and interact only in strictly prescribed ways. A mind-bogglingly complex network of interactions reduces to a few numbers that we call "the position" and "the time." The underlying complexity never goes away, though. In situations such as black holes, the system can become disordered and events can cease to have a position or a time. And even when the system is spatial, it contains a vast amount of latent complexity. The universe we see playing out in space may be just the surface level, where we float like little boats while leviathans stir in the deep.

Networking Space

The concept of space as a network goes back to the 1960s and the brainstorms of such innovative (and iconoclastic) theorists as John Wheeler, David Bohm, Roger Penrose, and David Finkelstein. Wheeler, for one, imagined taking a bucket of "dust" or "rings"—primitive grains of matter that do not exist within space, but simply *exist*—and stringing them together to form space. Physicists have been trying to make the idea work for decades. Today one of its strongest champions is Fotini Markopoulou, who pictures the stringing-together process as a graph akin to those Facebook diagrams. She and her colleagues call their approach "quantum graphity"—cutesy, but any effort to inject a sense of humor into physics jargon has got to be a good thing.

Quantum graphity doesn't specify what the Wheelerian grains actually are—that's a job for a full-up theory of quantum gravity such as loop quantum gravity or string theory. Quantum graphity is a theory-in-miniature that focuses narrowly on what you might build with those grains. Indeed, Markopoulou and her colleagues' philosophy is that the detailed composition shouldn't matter; the principles of organization should be universal. After all, physicists have found that similar rules govern a huge diversity of complex systems, from earthquakes to ecosystems to economies. On the downside, quantum graphity is so bare-bones that it faces the problem of meshing with known physics. "Fotini tries to jump straight in, but it's very ambitious and dangerous, because you have no connection to existing theories," says Claus Kiefer of the University of Cologne.

The links between the elementary grains are as simple as can be. Two grains are either connected to each other or not, like Facebook users who can either be friends or not—just an on-or-off relationship. The resulting network looks like a string-art craft project in which you hammer nails (representing the grains) into a sheet of wood and stretch threads (the links) among some of them. Despite the simplicity of its construction, the network can take on a huge variety of shapes, ranging from skeletal outlines to elaborate mandalas.

To breathe life into the network—to give it the capacity to transform and evolve—Markopoulou and her colleagues suppose that the links switch on or off depending on the amount of available energy. This process is ad hoc, but again the goal is not to create a bulletproof theory, but to reconnoiter possibilities for how to construct space. Each link represents a certain amount of energy. Chains of links contain less energy than an equivalent number of isolated links, so the total energy of a network depends not only on the sheer number of links, but also on how they're put together. The more intricate the pattern is, the more energy it embodies.

The energy maxes out in a fully interconnected network, where every grain is linked to every other grain. In such a network, the principle of locality doesn't hold; you can go from any grain to any other grain in one hop, without passing through any intermediate points. The network lacks the hierarchy of relations—near versus far, small versus big—which is characteristic of space. You can't subdivide it into

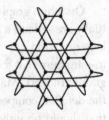

6.4. Quantum graphity model of space. In a high-energy network (*left*), every point is connected to every other point. In a low-energy network (*right*), every point is connected only to a few others, which become its nearest neighbors in the structure of space. (Courtesy of Fotini Markopoulou)

separate chunks; it's an indivisible whole. "This thing has no notion of locality . . . ," Markopoulou has explained. "If you just put out your hand, you reach everybody in the whole universe."

To see why the high-energy network is not spatial, try assigning locations to the grains. Every grain has to be equidistant (a single hop) from every other. For the first three grains, that's no problem: arrange them in an equilateral triangle. Four can be stacked in a pyramid. But where does a fifth go? There's nowhere equidistant to the first four points, at least not within ordinary, three-dimensional space. You need a *four*-dimensional pyramid. In fact, each additional grain requires a whole new dimension of space. Before long, you enter an ultra-higher-dimensional realm beyond our capacity to visualize. And most of that vast venue is wasted: the network is only one hop wide in any direction and does a good impression of a balled-up spider's web. So although you might still talk of the network as existing within space, it's not the kind of space we want: three dimensions that extend as far as we can see in every direction and provide an economical description of the relations among objects.

Lower-energy patterns are a different story. They're just what we want. Each grain connects to just a few others, forming a regular grid like a honeycomb or woven fabric. The notion of distance regains meaning: some grains are close together, the rest far apart. The network is nice and roomy. The principle of locality holds: for an influence to go from one place to another, it can't hop straight there, but must work its way through the network. The passage of the signal takes time, which would explain why the speed of objects through space is limited (by the speed of light).

In short, spacelessness and space are just two different phases of the same network of grains. One can metamorphose into the other; a crumpled wad can unfurl into a flat expanse. Theorists have proposed a couple of ways this might happen. The reshaping could be a process that occurs in time. The network starts off as sizzlingly hot—a highly interconnected pattern containing an enormous amount of energy. Then it cools off and crystallizes like a tray of water freezing to ice, as links dissolve and reorganize to create a tidy arrangement. The trick is to explain the cooling. Things don't just cool down on their own; something must drain them of heat. "Where did the energy go?" wonders Markopoulou. "You need a freezer. You need to cool the universe." She and her colleagues speculate that the energy could go into the creation of matter. The primordial grains could coalesce into elementary particles, so that matter emerges hand in hand with space.

Alternatively, the transition may not be a process that unfolds in time, but a structuring that arises at the quantum level. The network can exist in multiple conditions at once, a limbo known as superposition. Although most of those conditions are nonspatial, they can fuse together into something that *is* spatial. The best-developed account of superposition of space goes by the somewhat unwieldy name of "causal dynamical triangulations." Its inventors have shown that nonspatial geometries neutralize one another, as long as events are highly ordered, with a distinction between cause and effect built in from the outset. The effect is analogous to the wisdom of crowds, those remarkable cases in which you pose a question to a group and no one person has the right answer, yet pooling everyone's guesses *does* give the right answer. The classic example is the jelly-bean experiment: if you ask a group of people how many jelly beans there are in a jar, the average of their estimates will be better than any single person's estimate. The group has a collective intelligence beyond that of its members.

To delve into how the superposition might work, let's revisit Schrödinger's infamous cat scenario and take it a step further. In the original scenario, an experimenter puts a furry kitten into a limbo of being both alive and dead at once. A fusion of familiar conditions (life and death) leads to an unfamiliar one (the cat's existential ambiguity). For most people, that's weird enough, but it gets better, because the converse is also true. A fusion of *unfamiliar* conditions can lead to a *familiar* one. That means an indubitably alive cat is in a superposition, too—a superposition that fuses exotic conditions such as being "somewhat alive and somewhat dead." The "somewhat dead" parts cancel out, leaving an entirely alive mammal—much as, in the jelly-bean experiment, individual people's errors offset one another. Even an ordinary cat, in other words, is a bundle of contradictions (not that this will come as any surprise to cat owners).

Weird though this sounds, it's where the true power of quantum mechanics lies. Much of the everyday world is the result of bizarre processes whose bizarre aspects negate one another. For instance, why does light travel in straight lines? If you look at the quantum level, light actually moves on every possible path at once, but the twisty paths negate one another, leaving only a straight one. Space might be like that. The ordinary space we inhabit could be a superposition of nonspatial networks, the nonspatiality of one negating the nonspatiality of another.

Inside the Matrix

String theorists have explored ideas similar to Markopoulou's quantum graphity. In the 1990s they pioneered "matrix models," so called because the equations are based on grids, or matrices, of numbers much like the mileage charts. A matrix in the mathematical sense has nothing to do with the virtual-reality "matrix" of the movie *The Matrix*, yet the premise is eerily similar: the world we experience is a kind of simulation generated by a deeper level of reality. The best-known matrix model was developed by a quartet of theorists, Tom Banks, Willy Fischler, Steve Shenker, and Leonard Susskind. Their model, like quantum graphity, supposes that the universe is a cat's cradle of interconnections among grains of primitive matter. Under the right conditions, extraneous connections rupture and the grains snap into a regular spatial grid. "You start with a bunch of Tinkertoy parts with no particular structure to them," says Susskind, who is a physics professor at Stanford University. "You shake it up and it emerges into a lattice or structure of some sort."

String theory outgrew its name long ago. It postulates not just one-dimensional strings, but also two-dimensional membranes and higher-dimensional analogues—as theorists call them, 1-branes, 2-branes, 3-branes, 4-branes, and so on. Some branes, designated by D, can act as the endpoints of strings. At the bottom of this pecking order is the humble D0-brane, a type of particle. Being a true geometric point lacking size or any other spatial attribute, the D0-brane is the perfect building block for space. Confirming this intuition, theorists calculate that the D0-brane has the right properties to serve as the graviton, the particle that has been hypothesized for decades to convey the force of gravity.

Matrix models take this particle as fundamental and construct the universe entirely from lots of them. Every particle can interact with every particle, and their interactions are not simply on or off, but can vary in strength and in quality. The more energy you inject into a pair of particles, the tighter their bond will become. The namesake matrix of numbers quantifies this web of interactions. For example, if you read down to the eighth row and then across to the twelfth column, the number there will tell you how strongly particle number eight interacts with particle number twelve. To express not just the raw strength but also the quality of the connection, you need several such matrices.

Each matrix is a square, and running diagonally from the top left corner to the bottom right is a special set of numbers—where the eighth row meets the eighth column, the twelfth row meets the twelfth column, and so on. These tell you how much each particle interacts with *itself*. Self-interactions are a core feature of matrix models. The particles are subatomic narcissists, the physics equivalent of Facebook users who always "Like" their own posts. Their self-interactions have a carefree, unrestrained quality; you can dial their strength up or down without having to pump in energy.

Whereas the workings of quantum graphity are somewhat ad hoc, the laws governing D0-branes are dictated by considerations of symmetry. The mathematical balance of the equations is the organizing principle of this model. Symmetry ensures that the off-diagonal values in the matrix are yoked to the diagonal values—in other words, that the branes' mutual interactions depend on their self-interactions. Particles that self-interact by comparable amounts forge a bond, whereas particles with differing levels of self-interaction remain aloof. Put simply, like attracts like. Consequently, the branes agglomerate into separate clusters like the social circles in your Facebook network. These clusters constitute the ordinary subatomic particles of physics. Each cluster can be compactly described by a few numbers—namely, the strength and quality of its constituents' self-interactions.

That's how space arises in matrix models. The D0-branes don't live or move within space. Mathematically they all sit on top of one another at a single point. But because they're so selective about their interactions, they produce our experience of living within space. What we call "position" is simply the set of numbers that uniquely identifies a given cluster. It's like pigeonholing your friends as "physics lovers," "Radiohead groupies," or "Cuban-style dancers."

That's just the start. You can take all our familiar spatial notions—movement, size, locality—and explain them in terms of brane dynamics. Movement: things shift their position because the D0-branes' self-interactions are varying. It's like saying the Cuban dancers suddenly all get interested in Dominican music. They "move" as a group to a new passion. Such movement may sound metaphorical, but in matrix models it's the origin of physical movement. Size: the self-interactions of the branes in an object are not exactly equal, but have a slight spread, so that the object spans a range of positions. Locality: clusters at separate locations are independent because their self-interactions differ, which suppresses their mutual interactions according to the logic of symmetry. This is like saying that Cuban dancers and Radiohead groupies never have much to say to each other. "Things that are 'separated' are not really separated," Susskind explains. "There's just a cancellation of the things that are connecting them."

If all the branes did was reproduce space, that'd be gratifying, but boring. Our goal is to go *beyond* space. Branes can do that. They can behave in ways that are too complicated to represent with a handful of spatial coordinates. For instance, the mutual interactions among clusters are never fully suppressed, because quantum effects keep tickling them back to life. Therefore, spatially separated clusters are not fully independent; they feel the gentle tug of other clusters. This is how matrix models explain the force of gravity. In a sense, the models evoke Newton's picture of gravity as a nonlocal force that leaps from one thing to another. The interactions that produce it aren't transmitted through space, but are direct, unmediated, nonlocal links.

Another departure from spatiality occurs inside clusters. The internal group dynamics are intense and every brane is interacting with every other. The branes scramble one another's self-interactions, and the matrix values representing those interactions lose the qualities of spatial coordinates. Ordinarily, coordinates are independent numbers: you can measure the latitude of a city separately from its longitude. But you can't do that for branes within a cluster. If you measure the latitude of a brane first, then its longitude, you might get a different result than if you measured the longitude, then the latitude. This kind of ordering effect is known mathematically as "noncommutativity." In effect, the particle seems to be located in two different places, like Salt Lake City in my cities example. "The position in, say, the 'x' direction and the position in the 'y' direction can't simultaneously be measured," says Emil Martinec, a string theorist at the University of Chicago. "This is certainly not the behavior we expect for a collection of discrete particles—we expect to be able to localize them precisely in all spatial dimensions." The degree of ambiguity is a measure of just how nonlocal and nonspatial the system is.

Indeed, the cluster doesn't really have an "inside"—there is no volume of space where the D0-branes bustle around. Arguably there aren't even any D0-branes anymore, either, because they surrender their individuality and become assimilated into the collective. If you look at a cluster from the outside, what you see isn't the outer surface of a material thing, but the end of space; and if you poke your hand into the cluster, you will not reach into its interior, for the cluster has no interior. Instead, your hand will become assimilated, too (which can't be good for it). If you wisely refrain from touching the cluster and instead throw particles into it, you will notice that the cluster's storage capacity depends on its area rather than on its interior volume—again, for the simple reason that it doesn't actually have an interior volume. Space has no meaning at this level.

Holographic Reality

Matrix models do have some peculiarities, but they establish a remarkable principle: a bunch of particles obeying quantum physics can organize themselves so that you'd swear they live and move within space, even if space wasn't in the original specification of the system. And it turns out that this principle is very general. Not just a swarm of D0-branes but almost any quantum system contains spatial dimensions folded inside it like a figure in a pop-up book. Most such systems don't bootstrap space from utter spacelessness, as matrix models do, but prime the pump with a low-dimensional space in order to generate a higher-dimensional one.

The AdS/CFT duality that I mentioned in the previous chapter is such a system. It starts with a three-dimensional space and generates a nine-dimensional one. One reason string theorists like this scenario so much is that it neatly explains the holographic principle, the idea that the universe can sustain much less complexity than the principle of locality would lead you to expect. The complexity is reduced by just the amount you'd expect if one of the dimensions of space were illusory. In the AdS/CFT scenario, that's because the dimension in question *is* illusory. It can be collapsed down like an accordion because it was never really there. ("Illusory" is perhaps the wrong word. "Derived" or "constructed" would be better, if less poetic. The dimension may not exist at the lowest level, but it is still very real to anything larger than a brane.)

The disposable dimension reflects a particular aspect of order in the underlying quantum system. In fact, the requisite order is familiar to us from everyday life—specifically, the fact that big things and small things live as if in worlds apart. Our planet trundles around its orbit oblivious to human affairs, just as we spare little thought for the bacteria that lodge in our skin. Conversely, we have only a vague awareness of riding on a giant ball of rock, and bacteria know nothing of our daily struggles. Nature is stratified by scale.

Sound waves are an especially simple example of this stratification. Sounds of long and short wavelengths are oblivious to each other; if you sound a deep bass note and a high treble pitch simultaneously, each ripples through the room as though it were the only sound in the world. Their mutual independence is analogous to the autonomy of spatially separated objects. Suppose you play two piano keys, middle C and the adjoining D key. The C key creates a sound wave with a wavelength of 1 meter 32 centimeters, and D produces one with a wavelength 14 centimeters shorter. These waves overlap in the three dimensions of space through which they propagate, yet they're independent of each other, as if they were located in different places. In a sense, you can think of the sound waves as residing 14 centimeters apart within a fourth spatial dimension.

The farther apart the keys are on a piano keyboard, the farther apart they are within this imaginary dimension; a given distance along the keyboard translates into a given distance within the dimension. You don't see this dimension as such; to you, it's an abstraction that captures the acoustical independence of sound waves. But it's a remarkably fitting abstraction. Musicians call the difference between pitches a musical "interval," which has connotations of distance, as if our brains really do think of the differences between pitches as spatial separation. AdS/CFT duality takes this abstraction literally and suggests that one of the dimensions of the space we occupy represents the energy or, equivalently, the size of waves within the underlying system.

Raman Sundrum, a string theorist at the University of Maryland, has a dramatic way of putting it. Suppose you're an artist painting the National Mall, with an ice-cream stand in the foreground and the Washington Monument in the background. To evoke a sense of distance on the flat canvas, you draw these two objects at different scales. Something like that is happening for real in the AdS/CFT scenario. The universe looks three-dimensional, but could really be a two-dimensional canvas, and what we perceive as distance along the third dimension is ultimately a difference in scale. "The depth dimension could be recreated in the way that artists have to do it: by just drawing the Washington Monument really small and drawing something in the foreground really big," Sundrum says. A faraway object is actually sitting right next to you; it looks small because it really *is* small. You can't touch it not because it's distant but because it's so tiny that your fingers lack the finesse to manipulate it. When things grow or shrink, we perceive that as movement. It sounds fanciful but is backed up by rigorous mathematics.

Things of different sizes aren't strictly independent; they interact with things of comparable size, and the effects can cascade from one scale to the next. Consider the proverb of the nail: for want of a nail, the shoe was lost; for want of a shoe, the horse was lost; then the knight, the battle, and the kingdom. A nail shortage in a single blacksmith shop didn't immediately cause the monarch's downfall; it exerted its influence indirectly, via systems of intermediate scales. Sound waves of different pitches can also behave like this. A Chinese gong begins rumbling at a low pitch and gradually vibrates at successively higher pitches. The necessity of propagating through scale explains why spatial locality holds in the emergent dimension. What happens in one place doesn't jump to another without passing through the points in between.

It's not automatic that the underlying quantum system would possess this kind of hierarchical order. Just as a painting must be composed in just the right way to produce the sense of depth, so must the system have a certain degree of internal coherence to give rise to space. What ensures this cohesion is entanglement among the system's particles or fields. To produce space as we know it, those particles or fields must be entangled by scale: each particle with its neighbor, each pair of particles with another pair, each group with another group. Other patterns lead to different geometries or systems that can't be thought of as spatial at all. If the system is less than fully entangled, then the emergent space is disjointed, and an inhabitant of the universe would be trapped inside one region, unable to venture elsewhere. "Quantum entanglement is the thing that is responsible for connecting up the spacetime into one piece," says Mark Van Raamsdonk, a theorist at the University of British Columbia. When we first encountered entanglement, it seemed to transcend space. Today, physicists think it might be what *creates* space.

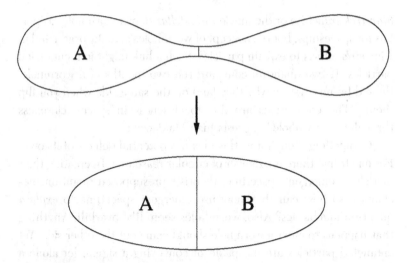

6.5. Connectivity of space. As two parts of the universe become increasingly entangled with each other, they draw closer together. (Courtesy of Mark Van Raamsdonk)

Glitches in the Matrix

To recap, we've seen several ways that space might emerge from spacelessness, all variants on the idea that a knotted network of primitive building blocks could straighten itself out into a tidy crystalline lattice. Now let's apply these tentative ideas to the nonlocal phenomena that I introduced in chapter 1, including the correlations between entangled particles, the grand-scale uniformity of the cosmos, and the fate of matter that falls into black holes. These phenomena may be glitches in the matrix: irregularities that betray the deep nature of reality.

The emergent spatial lattice isn't perfect—nothing in nature ever is. It has slight defects, like sewing mistakes that leave straggly threads running from your shirt collar to the cuff. Adjacency within the network need not imply adjacency within space; one small step in the network could be one giant leap in space. Markopoulou and Smolin call this phenomenon "disordered locality," while the string theorist Brian Swingle of Stanford University uses the term "long links." These links are essentially tiny wormholes, like the spacetime tunnels that general relativity theory permits. Science-fiction shows such as the

Star Trek franchise or the movie *Interstellar* depict wormholes as portals for spaceships, but the concept of wormholes actually originated in Einstein's efforts to explain particles. Such a link might join entangled particles. If two quantum coins are the two mouths of a wormhole, there'd be no mystery why they land on the same side when you flip them. "The correlations are due to closeness in space—closeness through the wormhole," suggests Juan Maldacena.

Compelling though it is, this idea has potential defects of its own. For one thing, there is a danger of circular reasoning. In creating their models of emergent spacetime, theorists presupposed quantum mechanics, so how can they now use emergent spacetime to *explain* quantum mechanics? Also, wormholes seem like overkill. Anything that happens to enter a wormhole should come out the other side. Yet entangled particles are incapable of conveying a signal, let alone a spaceship. Any tunnel between them would have to be a collapsed mineshaft, providing just enough of a connection to make the particles give matching results in laboratory experiments, but not enough to open up a general-purpose passageway. Proponents offer several ways to strike this balance. Maldacena argues that gravity pinches off the wormhole before a signal could make it through. Swingle suggests that experimenters would find it so inordinately difficult to stuff a signal through such a narrow link, requiring an implausibly powerful computer to encode and decode, that they could never do it in practice. And Markopoulou imagines that the wormhole *can* convey a signal, but the signal is meaningless because the particles are moving randomly— just as a cat walking on your computer keyboard might send an e-mail, but a gibberish one.

Others think entanglement demands more radical ideas. Michael Heller advocates a concept called noncommutative geometry, which is a slightly ironic name: the aim is actually to do away with geometry and describe the universe as a big algebraic equation. The equation is formulated in terms of matrices, like those of matrix models, but given a new interpretation. The matrices are no longer arrays of numbers describing the connections among building blocks such as D0-branes, but individual entities in their own right, the primary ingredients from which everything else is made.

In concrete terms, Heller and others pursuing this approach take a top-down view of physics, in which global structures—ones that span the entire universe—are fundamental, and local geometric concepts such as "points" and "things" derive from those global structures, rather than the usual bottom-up view in which the universe is built from zillions of localized things. "There are no points, no time instants," Heller says. "Everything is global. So locality has been engulfed by globality." By analogy, imagine that instead of defining society as millions of individuals who assemble into sundry groups, we define it as millions of *groups* and identify each individual as a bundle of group memberships. (This isn't so far-fetched: philosophers and sociologists such as Hegel have argued that individual identity is constructed largely from social identity.) You could draw a big Venn diagram of groups and every person would be the unique intersection of some set of circles. For this top-down definition to work, groups must overlap in just the right way. If they don't, individuals lose their distinct identity; you'd get more than one person with the same set of group identities and be unable to tell them apart.

Something like this could happen with the global structures of noncommutative geometry. Under the right conditions, these structures interlock to produce a spatial universe governed by the usual laws of physics. Under other conditions, they don't—and their imperfect meshing would produce nonlocal phenomena such as entangled particles. If two distinct objects are merely different mixes of the same global entities, it stands to reason that they can retain a connection that transcends space. What happens in one place will be sensitive to what happens elsewhere, even without any communication in an ordinary sense.

•

Emergent-spacetime models also give us a new way to understand the big bang. The genesis of the universe has always presented something of a paradox. Nothing is supposed to precede it, yet something must precede it to set the cosmos in motion. But the paradox dissipates when we think of the big bang not as an abrupt moment of creation, but as a transitional process. If space emerges from spaceless building blocks just as life emerges from lifeless atoms, then the birth of the universe is no more inscrutable than the birth of a living creature. The primeval world, form-less and void, gradually shaped itself into a spatial structure. The story recalls ancient creation myths in which primordial chaos becomes steadily more differentiated: earth from sky, night from day, sea from land.

Theorists toying with this idea have suggested several possible primordial states. According to quantum graphity, on the one hand, the universe was a network buzzing with energy. In effect, wormholes linked every point in space to every other. As the system cooled, the wormholes pinched off, different parts of the universe gained some autonomy, and space took on the form we recognize. Matrix models, on the other hand, suggest an alternative scenario. The D0-branes that constitute the universe at its deepest level were all interconnected, but that didn't translate directly to a linkage of points in space. In fact, space was splintered into countless disjointed pieces, an archipelago of little universes. Physicists call this condition "ultralocality." It is "ultra" in that the islands weren't just autonomous, but utterly isolated. Then, some of the D0-branes became entangled and started to behave as a single unit—a little patch of space. The newly formed unit, in turn, be-came entangled with other such units, creating a slightly larger patch of space. As regions became entangled on progressively larger scales, space rolled out like a carpet. From the point of view of someone living within space, it looked as though wormholes popped into existence and bridged the gap between islands.

Both scenarios would explain why the universe now appears so uniform. During the prebangian epoch, matter and energy would have sloshed around the network and spread out evenly before space germi-nated and grew. Two galaxies on opposite sides of the sky, at the far reaches of your vision, separated by a gulf of space, are unable to com-municate with each other *now*. But at the dawn of time, there was no space and no gulf between them. Because those galaxies were directly linked back then, their close resemblance is no coincidence. Only as space emerged did they sever their connection. "The idea came as a completely obvious consideration when I first played with a completely connected graph as a possible early phase of the universe," Markopoulou recalls. "You just look at it and go, duh, there would be no horizon problem here."

The emergence process might explain other general features of the universe, too. For instance, astronomers observe that the primordial universe wasn't exactly uniform, but slightly splotchy. These subtle deviations from perfect uniformity were the seeds around which galaxies and other cosmic structures coalesced later on. Intriguingly, the splotches look the same no matter what their size is. Such scale-independence is a common property of certain kinds of phase transitions—for instance, when high-pressure steam turns into liquid water; finely balanced between two states of matter, conditions fluctuate in a coordinated way on all scales. A phase transition from spacelessness to space could well produce the same pattern.

•

Black holes are like the big bang in reverse: the reversion of space to spacelessness. In the emergent-spacetime picture, they're puddles of water on the tundra: isolated spots in the universe where space literally melts, its orderly, crystalline structure degenerating into a fluid turmoil. We can scarcely even imagine what such a fluid is really like. "Whatever that is, it does not have a geometric interpretation," Martinec says. "The conventional geometric notion of spacetime has ended."

To the extent we can visualize the fluid at all, it is a highly interconnected network—not a gravitational vacuum cleaner so much as a cosmic labyrinth. It is effectively infinite-dimensional. Things falling in are unable to get back out not because gravity grips them firmly but because they have a hard time finding the exit. They might succeed someday, but in the short term they're goners. They suffer the fate of people who call their insurance company and spend all night navigating a Kafkaesque phone menu. "This highly connected area acts as a trap," Markopoulou says.

According to quantum graphity, the maze is literally a complex of passages, but in matrix models, it is more abstract—a labyrinth of complexity. The D0-branes that make up the hole are constantly in motion, always reshuffling themselves, as if wandering through a warren of possible arrangements. Occasionally a brane gets lucky: its connections to its partners momentarily slacken, so it flies off. Particle by particle,

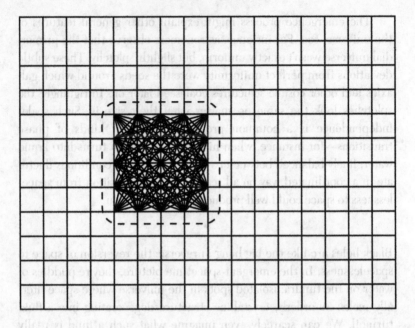

6.6. Quantum graphity model of a black hole. The black hole is a cosmic maze—anything that falls in can emerge only with difficulty. (Courtesy of Alioscia Hamma)

the black hole disperses, as Stephen Hawking predicted it should. "Getting trapped by the strong gravitational field for a long time before reemerging as Hawking radiation," Martinec says, "is largely related to the particle getting lost in this huge space of states."

Not only can the emergence of space account for nonlocal phenomena that physicists have observed or inferred, it predicts phenomena they've yet to see. One goes by the rather existentialist name of "bubble of nothing." Like a black hole, a bubble of nothing is a phase transition, but closer in spirit to boiling than to melting. A pot of water on your stove boils when bubbles of vapor nucleate, expand, and begin to interconnect, until the liquid is all gone. Likewise, a little pocket of spacelessness might nucleate and begin to grow. "At some point in time, quantum-mechanically a hole develops in spacetime," Rozali says. "The boundary of that hole propagates out at the speed of light.

It's sort of eating up spacetime." Happily, a bubble of nothing wouldn't suck you to your doom like a black hole; in fact, you could never see it directly, because it is outside space. Despite the name, the bubble is not strictly nothing. The quantum system hasn't gone away—just as water doesn't go away when it boils—but simply becomes too disordered to be thought of as spatial anymore.

•

Spaces might even be nested like matryoshka dolls. The space we live in might not only emerge from, but also give rise to, other spaces. Certain types of real-world systems behave as if extra dimensions of space are latent within them. For instance, materials near absolute zero can flow or conduct an electric current in unexpected ways, indicating that they've achieved an extra degree of organization. Their constituent particles live and move through three spatial dimensions but arrange themselves as if they're living and moving in *four* dimensions. If you look at these materials, you won't directly see the higher dimension. It exists at a different level of description in which clusters of particles, rather than particles themselves, are the primitive objects.

A similar thing may happen at the other end of the temperature scale. In the past decade, atom smashers such as the Large Hadron Collider have melted atomic nuclei to create obscenely hot states of matter. These boiling, roiling plasmas are an ungodly mess; the quarks and gluons in them interact so intensely that even the greatest math whizzes on the planet struggle to predict what they'll do. Yet many physicists think the plasma has an inner simplicity that becomes evident when you make the mental leap and think of it as existing in four dimensions. That higher-dimensional universe recombines elements of ours in almost unrecognizable ways. For instance, friction in the plasma can be reimagined as gravitational waves caroming off a black hole. In a sense, the Large Hadron Collider doesn't just create new forms of matter. It creates space.

•

It's astounding to think that space, thought for so long to be the rock-bottom foundation of physical reality, could perch atop an even deeper layer. Nonlocal phenomena are clues to this foundational structure, on the principle that you learn about how something is put together by noticing where it falls apart. Ironically, the main criticism I hear of quantum graphity, matrix models, and AdS/CFT isn't that they're too weird, but that they're not weird enough. All these models still work within the basic framework of quantum physics and general relativity, and much of the structure that is supposed to arise spontaneously is actually preprogrammed into the rules.

Notably, these models presuppose time; they don't incorporate Leibniz's and Mach's suggestion that time should emerge as surely as space does. Some researchers don't see this as a failing, but as a profound truth about nature—that time must be fundamental even if space isn't. After all, physics does need to have *some* foundational structure, something that everything else is built on, and time is as good a candidate as any. Indeed, how could you even talk about emergence as a temporal process if you don't presume time? "As soon as you say time is emergent, you run off the rails," Martinec says. "What are the rules? What do I do?" The Caltech cosmologist Sean Carroll has put it succinctly: "Space is totally overrated, whereas time is underappreciated . . . I think that time is going to last . . . Space, on the other hand—totally bogus. Space is just an approximation that we find useful in certain circumstances."

Yet this separation of time and space runs counter to Einstein's great insight that the two are fundamentally inseparable. If one is emergent, surely the other must be. Many physicists do think that time emerges and have been looking for ways to think of emergence without presupposing time. A clue comes from the holographic principle. So far, I've described the holographic principle as a way to generate space, but it can also generate time. The key in both cases is the existence of a boundary. If the universe has a boundary located out in deep space, the emergent dimension is spatial, but if it has a boundary in the past or future, the emergent dimension is temporal. In fact, as far as astronomers can tell, our universe has temporal rather than spatial boundaries. In the past, there's the big bang; in the future, eternally accelerating expansion, which also serves as a type of boundary. An observer sitting on that boundary in the distant past or far future would know all there is to know about the intervening moments. Yesterday, today, and tomorrow would collapse into one.

By this logic, theories that presume time are still incomplete, merely stepping stones to a complete account of how space and time emerge from deeper physics. Theorists will need an even more radical approach to explain nonlocality than those they've tried so far. And indeed they've been stumbling toward one.

Conclusion:
The Amplituhedron

Toward the beginning of the Second World War, Werner Heisenberg made a famous and historically controversial visit to Copenhagen, during which he and Niels Bohr discussed Heisenberg's efforts to build an atom bomb for the Nazis. Heisenberg took a lesser-known but equally dramatic trip toward the end of the war. In December 1944, he crossed from Germany into Switzerland to give a talk at the University of Zurich. Sitting in the audience, along with some of his old physicist friends, was a man he didn't recognize—from what he could tell, a local physics aficionado or perhaps an SS agent sent to make sure he wasn't bad-mouthing Hitler. In fact, it was Moe Berg: ex–baseball player, Princeton-educated linguist, and American spy. Berg had been sent to find out whether Heisenberg was close to finishing the bomb and, if so, to kill him. When Heisenberg stayed mum about his bomb work and lectured instead about a side interest, a new quantum concept called the S-matrix, Berg let him live.

The S-matrix was a revolutionary approach to doing physics without space and time, one that was even further removed from ordinary notions of space than the graphs and matrices that I talked about in the previous chapter. Heisenberg, who had always had it in for space, felt vindicated by the troubles quantum field theory was having in explaining electric and magnetic forces—specifically, its prediction that these forces should act with infinite strength. To sidestep the question of whether the theory was right and what might replace it, he devised a mathematical version of the principle that what you don't know can't hurt you.

He proposed to treat messy particle collisions as a black box. You know what goes in, you know what comes out, but no one ever sees the confused jumble in between. The S-matrix tabulates the probabilities of the sundry possible outcomes. To calculate its entries, Heisenberg argued, theorists don't need to know what goes on inside the box. They can neglect where particles are, how they move, and even whether they really are particles, as opposed to ripples in a field or some weird thing that physicists have yet to imagine. In short, theorists can banish any mention of the concept of space from their description of physics. Instead, they can deduce what they'll observe based on broad rules. It's like rolling dice. You could use a supercomputer to solve the equations of motion for small dimpled cubes that are tumbling through roiling air currents. Or you could take a shortcut: because of symmetry, a die has an equal chance of landing on any of its six sides.

Fortunately for us all, Heisenberg's mathematical contraption worked rather better than his bomb ever did. The S-matrix became a part of every theorist's toolkit. But not for the reasons Heisenberg originally thought: it was considered a handy accounting system rather than a way to do without space and time. Not long after the war ended, physicists figured out how to use quantum field theory to do full-up calculations—to open up the box and look inside—and put aside the question of whether space and time ultimately disintegrate. Yet the black box slammed shut once again in the 1950s and '60s as physicists plumbed the depths of atomic nuclei. Quantum field theory didn't seem up to the task of describing nuclear forces, and the S-matrix regained its appeal. This time, the theorist Geoffrey Chew of the University of California, Berkeley, took an extra step. Whereas Heisenberg had assumed there were some underlying laws of physics—a mechanism operating within the box—Chew suggested there weren't. Maybe the S-matrix is all there is.

This was radical, and radicalism went over well in '60s-era Berkeley. Part of Chew's goal was to get rid of space and time, which he held responsible for the failings of quantum field theory. "To make major

progress we must stop thinking and talking about such an unobservable continuum," he told his colleagues in a lecture in 1963. Instead of
a step-by-step narrative of particles or waves propagating through space,
the laws of physics, Chew proposed, were a set of principles for how
things or processes relate to one another. The innards of the box aren't a
clockwork of moving parts, but a puzzle that fits together in a certain
way. Not only are the parts not moving, they aren't really "parts." Within
the atomic nucleus, nothing is more fundamental than anything else;
everything has its place in the structure. The S-matrix describes this
structure mathematically, and physicists could approach it like sudoku:
fill in the grid of numbers based on simple rules. The space and time
we perceive on macroscopic scales derive from the subatomic order.

Yet the program floundered. It predicted that basic principles
would fully determine the S-matrix. Chew wrote: "Nature is as it is because this is the only possible nature consistent with itself." But in fact
there's no unique S-matrix for the particles Chew was studying. General rules don't tell you where to put all the numbers, just as a poorly
crafted sudoku doesn't give you enough information to complete it. By
the early 1970s, quantum field theory proved able to explain the nuclear forces the old-fashioned spatiotemporal way, and for a second
time most physicists left the S-matrix for dead.

•

As S-matrix theory was riding its roller coaster of fortune, the Oxford
mathematician Roger Penrose was working on his own way to think of
spacetime as emergent. He initially tried to conceive of space as a network, as I talked about in the previous chapter, but he found he could
account for only some aspects of space, so in the 1960s he expanded
the framework into something he called "twistor" theory. Inspired by
quantum nonlocality, Penrose reasoned that nonlocal structures had to
be more fundamental than local ones. So he built his new theory not
on particles or other localized building blocks, but on light rays. Penrose wasn't interested in light per se—as a means of illumination—but
in the causal links that light rays represent. Light rays stretch infinitely
far across space, so they're about as nonlocal as you can get. Out of
them you can construct all the conventional structures of physics.
The intersection of light rays gives you a point. A swirling pattern of

rays—the origin of the word "twistor"—reproduces a spinning parti-cle. "Local structures in spacetime are encoded nonlocally," says Lio-nel Mason, one of Penrose's former students and now a colleague of his at Oxford.

As strange as it might seem to take a light ray as a basic unit, it hews closely to how we perceive the world. We never observe spacetime points and distances as such; what we do observe are light rays. Another of Penrose's colleagues, Andrew Hodges, says: "The twistor picture is much closer to the way we think . . . Line-of-sight ideas are a very fundamen-tal thing. We don't have a direct intuition of spacetime events." The next time you go out to look at the night sky, consider that you are connected by light rays to all those stars. In a sense, you're closer to those stars than to someone sitting right next to you at the same instant, because light is reaching you now from the stars, whereas there's a slight lag between you and that person. The spatiotemporal distance, defined by relativity theory, between you and those stars is zero. As Rafael Sorkin, who has worked on similar ideas, puts it, "A star is closer than yesterday."

Alas, the approach faded out. A major difficulty was what Pen-rose called the "googly" problem (adopting a term from cricket) or what a *Buffy the Vampire Slayer* fan might call the vampire prob-lem: the particles of the theory did not cast an image in a mirror. "He's very dogged, but somehow it just wasn't working," Mason says. Most physicists thought Penrose was wasting his time. "Twistor theory was the ugly duckling of theoretical physics," Nima Arkani-Hamed has re-called. Penrose didn't help his cause with his outspoken skepticism of competing approaches, notably string theory, which has gone through its own cycles of boom and bust. Valid though his critiques might have been, they weren't calculated to endear him to his colleagues.

Over and over, you see the same pattern: a brilliant idea about space has a meteoric rise and fall. Some people have concluded that the time just isn't ripe for getting answers, and they have to weigh the fascination of the topic against its frustrations. Hans Halvorson, for in-stance, gave up thinking about nonlocality to delve into other areas of philosophy. "Anyone who has thought about this goes through phases of excitement and depression," he says. "I'm feeling a bit depressed now." Fotini Markopoulou has oscillated between exhilaration and exasperation for years. "I'm discouraged more than I'm encouraged," she told me over brunch in 2011. "I wonder whether I should spend my life doing this. It's not like you see a lot of results." When I caught up with her again a year later, she had left science to study industrial design. "I do absolutely think that the questions of quantum gravity need to be answered, but you can't really pull an answer out of thin air . . . It's the rest of my life. I'd really like to have some time to experiment a little bit."

•

In the fall of 2003 Penrose visited Princeton University and gave a series of lectures in which he deemed string theory a "fashion." He recalls feeling apprehensive about meeting Edward Witten, the leading string theorist, as well he might. But Witten hadn't even gone to the lectures, and he was much more interested in picking Penrose's brain about some new ideas. "He started explaining something to me," Penrose remembers. "It looked awfully like it had to do with twistors." Witten asked Penrose to look at a short article he was writing about Penrose's old brainchild. The "short" article turned out to be a seventy-page monster that tied strings and twistors together in a single theory. "I found it very fascinating and exciting," Penrose says.

Witten has long been a crossover figure—not just a leader of string theory, but a point of contact with other fields of research, notably pure mathematics. He had even written a paper on twistors in their heyday twenty-five years earlier. "I had been extremely interested in twistor theory ever since I first heard of it," he says. "I made many tries to do something useful with twistor theory without really getting what I wanted. I feel that I was basically thinking off and on about possible directions to use twistor theory for all those years."

Witten's paper became a case study in how physics continually reinvents itself, often by reaching across old boundaries. It provoked some soul-searching. Twistorians were astounded that string theorists, of all people, had answers to the googly problem and other issues that had vexed them. Mason began to realize how often they'd missed a chance to exchange ideas. For example, while visiting Syracuse University in 1987, he and Penrose blew off a talk by the particle theorist

Parameswaran Nair of the City College of New York, whose work presaged Witten's and, in hindsight, would have filled in the gaps in twistor theory. "We never met up," Mason says. "So this beautiful idea lay there for sixteen or seventeen years."

String theorists, too, were jolted to action. Witten's paper reconciled them not only to twistors, but also to an overlooked question within their own area of study: Why are the outcomes of particle collisions so darned hard to calculate, and is there an easier way? As I mentioned in chapter 1, most physicists saw such calculations as homework from hell they'd just as soon forget about. Only a few people such as Zvi Bern were doing much about it, and even they were stalling out by 2003. Following up Witten's paper, Freddy Cachazo of the Perimeter Institute and several colleagues proposed a way to do those calculations without spacetime coordinates—putting aside the mechanism of particle collisions and focusing on their inputs and outputs. The idea sounded eerily like the old S-matrix program, which now seemed to be rising from the dead all over again. "This history might be termed the 'Revenge of the Analytic S-Matrix,'" says Lance Dixon of the SLAC National Accelerator Laboratory, one of Bern's closest collaborators.

By the time Witten brought the string and twistor communities together, they'd worked in mutual isolation for so long that they barely spoke each other's language. Arkani-Hamed describes a peculiar exchange at a meeting in Oxford in 2005. Cachazo gave a talk about the new calculation techniques, and during the question-and-answer period, Hodges made a remark that nobody could really follow—something to the effect that Cachazo's S-matrix diagrams looked uncannily like twistor diagrams. "I just didn't get anything about this, zero, zilch," Arkani-Hamed recalls. "I thought this guy was either a complete crackpot or a total genius." Still, Arkani-Hamed was intrigued enough to do his own calculations and express them graphically. "The pictures started vaguely looking like Andrew's pictures," he says. He had no idea what the pictures actually *meant*, but they matched, and that had to be significant.

What Hodges had seen was a way to visualize the S-matrix calculations geometrically using twistors. In 2013 Arkani-Hamed and his graduate student Jaroslav Trnka, now at Caltech, unveiled a geometric

technique for calculating probabilities, known in the jargon as "amplitudes," for particle processes. They gave their technique a suitably
funky name: the "amplituhedron." Based on the particles involved in a
given process, you draw a polyhedron with one vertex for each particle.
For instance, if you have two incoming particles that create four outgoing particles, you need a total of six vertices—a hexagon or one of its
higher-dimensional counterparts. The momentum of a particle sets the
size of its corresponding polyhedral face. Having formed this shape,
you then calculate its interior volume, and that quantity, by the rules of
the procedure, equals the desired amplitude.

The polyhedron isn't a real object sitting in ordinary space but an
abstract mathematical shape that captures the structure of particles'
interactions. It subsumes all the previous calculation techniques that
physicists have used to compute amplitudes, including Richard Feynman's baroque diagrams and Bern and his colleagues' minimalist alternative. These different techniques correspond to different ways to carve
up the polyhedron for purposes of calculating its volume. The polyhedron also exhibits symmetries that nature possesses, but which theorists
had never glimpsed before.

This procedure doesn't presume that the process plays out in spacetime. "There are no fields, no particles, no interactions," Trnka says.
The locality we observe in daily life is a consequence of the way the
faces fit together—specifically, that they form a closed shape, as opposed to disconnected planes. Those six vertices link into a hexagon
rather than, say, an asterisk. In general, the faces *won't* fit together;
locality is therefore a special case. "Simple geometric properties of the
amplituhedron encode locality," Arkani-Hamed explains.

The main lesson, as with the other approaches to emergent spacetime, is that space represents a type of order in the world, one that you
might not expect a priori. To be sure, the technique so far works only
for highly idealized theories of nuclear forces, and the researchers still
have to extend it to the messier reality we live in. Moreover, they and
philosophers still need to fill in the physical interpretation of the structure. For now, they're better at describing what nature *isn't* than what
it is. "The building blocks don't have a spacetime interpretation,"
Arkani-Hamed says. "These building blocks come from a very different

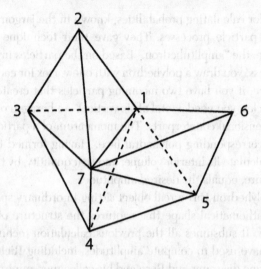

7.1. A portion of the amplituhedron that corresponds to the interaction of seven particles. (Courtesy of Jaroslav Trnka)

world than we think about in particle physics." What they actually are is another question. The histories of the future will probably straighten out all the kinks in the story and just present whatever the answer happens to be. Those of us who are living through this era know how much more interesting the search has really been.

Like Teenagers

So now we've seen two contemporary disputes over nonlocality. The twistor-string wrangle has unwound, while the entanglement debate I discussed in chapter 4 still rages (though, as I argued there, becoming moot, because all the positions suggest our notions of space have to give way). In neither case has the disagreement been resolved by the outright victory of one side over the other. Such clarity is rare in science.

If science is like a tree, most nonscientists focus on the trunk—the towering accumulation of knowledge. But that's the dead part of the tree. For scientists, the real tree is the thin layer of living tissue under the bark where the organism grows. We look to science for answers, but scientists are driven by questions, and to them an answer is merely a prelude to another question. We expect scientists to speak as one, but the very idea of speaking as one is alien to science.

Scientists tend not to discuss this aspect of their profession. They often describe their disputes as dirty laundry best not aired in public (and complain when journalists do so). But if you don't air disputes, what's left to air? Science *is* dispute. Once scientists reach consensus, they move on to some new argument. It's like having a dinner party with New Yorkers: if they manage to agree on anything, the moment of peace lasts just long enough to take another sip of craft beer. The field takes the kind of personality that values the process as much as the end result. "When you yourself discover something, you have a more intimate relationship with the ideas," Arkani-Hamed tells me. "There are all the blind alleys. You've seen it from all its unflattering angles. It's easier to love it; it's easier to know when to hate it."

Like scholars in general, scientists are judged on their originality, not their diplomacy. Their professional life is not unlike the social life of teenagers. It's insecure. They're always having to come up with ideas and maneuver to get them a hearing. The surest route to approbation is to show that their elders are doing something wrong and they can do it better. They make a name for themselves by taking a stand on a topic their peers consider important and, when challenged, doubling down rather than folding. In-group–out-group differences are strong; people establish a comfort level within their group that lets them share crazy ideas or ask potentially stupid questions, at the price of making disagreements with other groups loom larger than they probably are.

As frustrating as it can be for individual researchers to cope with stubborn colleagues, science as a whole is served by avoiding premature consensus. Every new idea deserves a forceful advocate. Sometimes a theory prevails despite what the evidence initially suggests, and sometimes it is fruitful even if utterly wrong. Plenty of the greatest advances have their origins in wrong ideas. There's an irony here. Scientists who are actively involved in a debate need to be convinced that they're right and anyone who disagrees must be missing something. Those of us watching from the sidelines know better, but we still want all the protagonists to *think* they're right, so that no one will give up too soon. We need to be tolerant of intolerance.

What breathes life into ideas, at least in the short term, is not whether they are right or wrong, but whether they get other people thinking. The success of any scientific or artistic creation is decided less by what it is than by what it provokes. Truth is the judgment of posterity; until then, fertility matters more. "Good ideas have a kind of Darwinian survival value that bad ideas don't," Leonard Susskind has written. "Good ideas tend to produce more good ideas—bad ones tend to lead nowhere." And an idea is more likely to propagate when it helps other people advance *their* ideas. It is a running joke among scholars that when they give a lecture and a hand shoots up, the questioner will ask some variant of: "That's very interesting, but how does it relate to my work?" A big reason why novel ideas take time to be accepted is that, by their very nature, they *don't* relate to other people's work.

Consensus emerges as the old questions lose the fertility that Susskind spoke of and new ones take over the ecosystem. Yesterday's drag-out fights are today's homework problems. But that takes time. On occasion, scientists do try to hurry up the process, but usually only in response to a request from the outside. Funding agencies ask advice on which projects to pay for; Congress seeks expert opinion on issues of the day; a science writer asks quantum physicists about nonlocality. These requests are like collapsing the quantum wavefunction. When a particle exists in some ambiguous state, the act of observing it can force the outcome, at the price of creating paradoxes. Likewise, insisting on consensus from scientists can have unwanted consequences. Consider all the fluctuating advice we get about what foods to eat or cancer screening to perform. That's what happens when we ask scientists for definite results before their time.

•

As ethereal as physics can be, its concepts have a funny way of entering into everyday life and, in turn, being influenced by broader cultural trends. "Our civilization forms an organic whole," Erwin Schrödinger said in a lecture in 1932. "Those fortunate individuals who can devote their lives to the profession of scientific research are not merely botanists or physicists or chemists, as the case may be. They are men and they are children of their age."

Although physicists often anticipate social trends, they have, in a way, fallen behind when it comes to concepts of space. It may seem radical to them that spatial distance could be a kind of illusion, but we're already living through "the death of distance," as the *Economist* writer Frances Cairncross has put it. Modern communications technology may not technically be nonlocal, but it sure feels that it is. The telegraph and telephone struck people in the nineteenth century as almost magical and made them rethink the boundaries of the self; we now hear much the same rhetoric about any new Apple product. And the information age is just the latest leg of a longer historical journey that began with the printing press and the oceangoing ship. A poignant example of how our concepts of distances have been utterly transformed came during Hurricane Katrina, when many evacuees said it was the first time they had ever left New Orleans. This immobility would hardly have been commented on a century ago, when people on average lived within a few tens of miles of their birthplaces. Today college-educated people, at least, typically live in several different places over the course of their lives. Commercial transport can take you to almost any human settlement on the planet within twenty-four hours.

Geography is no longer destiny. Our identities are still shaped by the old geography-based pigeonholes of nationality, ethnicity, and race, but we now have the option of constructing group identities for ourselves. We draw our circles of friends and colleagues based on shared interests and emotional intimacy, rather than on having been born in the same village. These webs of relations create their own notion of social distance that is distinct from spatial distance.

I've noticed that modern life has transformed even our direct sensory perception of actual physical distance. Eight hours a day of staring at a screen and accepting its illusory depth makes it harder to apprehend depth when we see it for real. When I was an astronomy teaching assistant in graduate school, students would look through the telescope eyepiece at Saturn, pull back as if they did not know what to make of it, look again, and ask: That's really Saturn? It's not a picture? A projection? Some students insisted on looking down the telescope tube from the other end to convince themselves they really were peering across 900 million miles of open space.

Ironically, though, as distance comes to matter less, it also seems to matter more. People complain that there is more daylight between us than ever before, that we have trouble connecting with one another as people, that the most we can say about our neighbors is that they seem so nice. We feel closer to those who are distant and farther from those who are close. Yet as much as we bemoan this trend, keeping one another at arm's length is the essential tradeoff of modern life. Love is meaningful because we remain individuals; if lovers' wish to meld into one were fulfilled, what would be left? On a grander scale, the greatest evils arise when distance closes up and we meld into a mob.

This ambivalence is now a defining element of physics. On the one hand, distance may not matter at a fundamental level. When two people are far apart, they might actually be right next to each other, considered in some deeper sense. On the other hand, our existence nonetheless requires the emergence of a concept of distance. For Einstein it was common sense that you must touch something to affect it, but we have come to realize how remarkable and fragile a fact this really is. "Space and time are the great uniters and dividers," wrote the early twentieth-century German philosopher Moritz Schlick. They make us individuals and relate us to one another. We can't have one without the other.

And if it does turn out that space and time are the products of some deeper level of reality, who knows what new phenomena await our discovery? Could cosmic mysteries such as dark matter and dark energy signify the breakdown of space? Might there be conditions under which we could travel faster than light (presumably in a way that forestalls paradoxes)? To me, these heady speculations pale beside a simple realization. If the ultimate constituents of the universe aren't spatial, they have no size, and they can't be probed by cracking matter into ever-smaller bits. They exist everywhere. They may well be right in front of our eyes and have gone unnoticed all this time. We may find the most exotic phenomena in the most prosaic places.

A Note on Entanglement

Half the fun of entanglement experiments is creating the entangled particles. In the apparatus I describe in Chapter 1, that task is performed by a crystal of barium borate. It looks like a little prism, but unlike your average glass prism, it does not passively transmit light, but can act as a catalyst for all sorts of novel optical effects. Ordinarily, when an electromagnetic wave passes through a material, the electric charges inside the material oscillate, and if a second wave passes through at the same time, the total oscillation is just the sum of the two individual oscillations. But in barium borate the combined oscillation is a complicated mixture. The crystal therefore lets two crossing light beams, or even just the different colors within a single beam, interact, which they wouldn't otherwise do.

The interaction can amplify one of the beams at the expense of the other. And the beam that gets amplified need not be a "beam": it could be just the residual quantum quivering of the electromagnetic field. The field spontaneously produces feeble flickers of light even in the absence of external illumination. If you set up the crystal properly, the amplification is so powerful that it turns this unpromising raw material into a proper light beam. Shine one beam on the crystal and two will emerge.

In a typical experiment, an incoming blue or ultraviolet beam conjures up two outgoing red beams. This process occurs particle by particle: each blue photon splits into two red ones. Because a red photon has half the energy of a blue, no energy is lost or gained. The splitting, known as spontaneous downconversion, is a low-probability event. The vast majority of photons in the incoming beam continue straight through the crystal unaffected; only perhaps one in 10 million interacts with the quantum fluctuations and divides.

By virtue of their common origin inside the crystal, the outgoing red photons have correlated properties: time of emission, energy, momentum, polarization. For simplicity, experimentalists usually concentrate on polarization—the direction in which the light waves oscillate (or, equivalently, in which the photons spin). The crystal is symmetric about an axis, and if an incoming photon is polarized parallel to that axis, the outgoing photons will be polarized perpendicular to it. Vertical polarization in, horizontal out; horizontal in, vertical out. Depending on how you arrange the beam and crystal, the outgoing photons can have the same polarization (known as Type I downconversion)—this is the setup in the experiment I focus on—or exactly the opposite polarization (Type II). Either way, their polarization has a fixed relationship guaranteed by the crystal's symmetry.

So, that gives you two matching photons. But for two *entangled* photons, you need an extra ingredient: the polarization has to be indeterminate. This is really the essence of entanglement, the reason it's so mysterious. The photons have the same polarization, but that polarization is not horizontal, not vertical, not diagonal, not circular. It's just plain nothing, a blank that has yet to be filled in, according to the standard interpretation of quantum mechanics.

After all, if the photons did have a specific polarization, there'd be no mystery in these experiments. You'd create two identical photons and later measure them to be identical, which is no weirder than pairing two socks as soon as they come out of the dryer and later observing that they're the same color. Entangled photons, in contrast, are like socks that don't have any particular color. If that boggles your brain, it should. This is what it means to be nonlocal: you can make a statement about the system as a whole—namely, "the parts are the same"—but not about any individual part. Each sock assumes a color only when observed, and both assume the same color, despite the apparent lack of any communication between them. How they do so is the puzzle of spooky action.

There are various ways to produce this indeterminacy with the barium borate crystal. In the version of the experiment I describe in Chapter 1, you sandwich two thin layers of the material, one oriented vertically, the other horizontally. Then you send in light that is polarized in neither a horizontal nor a vertical direction, but on a diagonal. As long as the layers are thinner than the beam, there's a quantum uncertainty about which layer the beam will interact with and, therefore, which polarization the outgoing photons will have. Their ambiguity is resolved only when they strike the polarizers and are measured. Usually in science, experimentalists do their utmost to eliminate uncertainty, but here they need to heighten it.

Just upstream of the crystal, you place an optical element known as a waveplate, which can be oriented to ensure the laser photons striking the crystal are diagonally polarized or not. This waveplate therefore decides whether the photons emerging from the crystal are entangled.

What's nice about the crystal is that it gives you a very controlled way to create entangled photons. But it's just one way. Before physicists discovered barium borate, they triggered atoms of calcium or mercury to release rapid-fire bursts of entangled photons, harvested entangled gamma rays from radioactive isotopes such as copper-64, and banged protons together to entangle them. Even the mercury vapor in ordinary fluorescent light bulbs emits entangled photons, albeit not in a readily visible way. Entanglement isn't a product just of shiny crystals. It's all around us.

Whatever the source of entangled particles, they pose the same conceptual puzzles, and to bring those puzzles into sharper focus, it's useful to think about entanglement at a higher level of abstraction, such as the coin-flipping metaphor I use through the book. But whenever entanglement seems too abstract, you can always relate it back to a concrete experiment.

Notes

INTRODUCTION: EINSTEIN'S CASTLE IN THE AIR

3 *"no previous discovery"*: Kafatos and Nadeau, *The Conscious Universe: Part and Whole in Modern Physical Theory*, 1.

3 *dating to the seventeenth century*: locality, *n.*, (1) The fact or quality of having a place, that is, of having position in space. 1628 Bp. J. Hall *Olde Relig.* vii. iii. 69 'It destroyes the truth of Christs humane bodie, in that it ascribes quantitie to it, without extension, without localitie.'" *Oxford English Dictionary Online*, accessed November 30, 2012, www.oed.com.

4 *famous essay in 1936*: Einstein, "Physics and Reality."

4 *"expect a chaotic world"*: Einstein, *Letters to Solovine: 1906–1955*, 117.

5 *mean different things*: Earman, "Locality, Nonlocality, and Action at a Distance: A Skeptical Review of Some Philosophical Dogmas."

5 *it had two aspects*: Howard, "Einstein on Locality and Separability."

7 *"a clean separation"*: Ibid., 187–88.

8 *the world without space*: Kant, *Critique of Pure Reason*, 50.

9 *he was consistently ahead*: Fine, *The Shaky Game: Einstein, Realism, and the Quantum Theory*, 16–25.

9 *Atoms in your body*: Smolin, *Life of the Cosmos*, 252.

9 *"spooky actions at a distance"*: Born and Einstein, *The Born-Einstein Letters 1916–1955: Friendship, Politics and Physics in Uncertain Times*, 155.

9 *sought a deeper theory*: Belousek, "Einstein's 1927 Unpublished Hidden-Variable Theory: Its Background, Context and Significance."

9 *"castle in the air"*: Stachel, *Einstein from 'B' to 'Z,'* 151.

10 *endow you with psychic powers*: Collins and Pinch, *Frames of Meaning: The Social Construction of Extraordinary Science*, chap. 4; Kaiser, *How the Hippies Saved Physics: Science, Counterculture, and the Quantum Revival*, chaps. 4, 10.

11 *not what we once thought*: Greene, *The Fabric of the Cosmos: Space, Time, and the Texture of Reality*, 123.

11 *space and time are doomed*: Gross, "Einstein and the Search for Unification," 2039.

11 *"single most astonishing discovery"*: Tim Maudlin, e-mail to author, October 17, 2012.

11 *"deeper, and more mysterious"*: Maudlin, "Part and Whole in Quantum Mechanics," 60.

11 *"What makes physics possible"*: William Unruh, interview by author, November 15, 2010, Utrecht, Netherlands.

1. THE MANY VARIETIES OF NONLOCALITY

13 *"playing with Erector sets"*: Enrique Galvez, e-mail to author, July 4, 2012.

14 *mad-scientist contraptions*: Gilder, *The Age of Entanglement*, chaps. 30–31.

14 *Apparently, clothes washers*: Markus Baden, interview by author, November 15, 2011, Singapore.

14 *"more fun than something exploding"*: Galvez, telephone interview by author, May 31, 2012.

14 *used entanglement to teleport*: Bouwmeester et al., "Experimental Quantum Teleportation."

15 *"just for the fun of it"*: Galvez, telephone interview by author.

15 *sweated over it for two years*: Holbrow, Galvez, and Parks, "Photon Quantum Mechanics and Beam Splitters."

15 *entangle particles in their basements*: Galvez, *Correlated-Photon Experiments for Undergraduate Labs*; Prutchi and Prutchi, *Exploring Quantum Physics Through Hands-On Projects*; Musser, "How to Build Your Own Quantum Entanglement Experiment, Part 2 (of 2)."

15 *"getting out to the masses"*: David Van Baak, interview by author, March 17, 2011, Dresden, Germany.

15 *in terms of metaphor*: Lightman, "Magic on the Mind: Physicists' Use of Metaphor."

15 *into two red beams*: Nikogosyan, "Beta Barium Borate (BBO)."

17 *"You would be surprised"*: Galvez, e-mail to author, October 8, 2012.

19 *stretched the distance*: Ursin et al., "Entanglement-Based Quantum Communication Over 144 Km."

19 *space-based version*: Morong, Ling, and Oi, "Quantum Optics for Space Platforms."

19 *close to real magic*: Mermin, "Is the Moon There When Nobody Looks? Reality and the Quantum Theory," 47.

20 *"Students love it"*: Galvez, interview by author, August 5, 2011, Hamilton, NY.

20 *article on the early entanglement experiments*: d'Espagnat, "The Quantum Theory and Reality."

20 *"My roommates remember"*: Maudlin, interview by author, January 19, 2011, Princeton, NJ.

20 *"he shut down the question"*: Maudlin, e-mail to author, October 17, 2012.

21 *championing or contesting*: Beller and Fine, "Bohr's Response to EPR," 23–27; Howard, "Revisiting the Einstein-Bohr Dialogue," 59, 81–82.

21 *"fame would be undiminished"*: Pais, *Einstein Lived Here*, 43.

21 *deemed them "philosophical"*: Kaiser, *How the Hippies Saved Physics: Science, Counterculture, and the Quantum Revival*, chap. 1.

21 *"satisfying description of nature"*: Dirac, "The Evolution of the Physicist's Picture of Nature," 48.

21 *"get to the bottom"*: Maudlin, interview by author.

22 *airing his misgivings*: Whitaker, "John Bell in Belfast: Early Years and Education," 14–17.

22 *was not cited*: Kaiser, *How the Hippies Saved Physics*, 319n41.

22 *One of his obituaries*: Gribbin, "The Man Who Proved Einstein Was Wrong"; Bell, *Speakable and Unspeakable in Quantum Mechanics*, 150.

22 *"an apparent incompatibility"*: Ibid., 172.

23 *"peaceful coexistence"*: Shimony, "Conceptual Foundations of Quantum Mechanics," 388.

23 *"no fundamental conflict"*: Maudlin, e-mail to author, October 18, 2012.

23 *not even the sneakiest government surveillance program*: Ekert, "Quantum Cryptography Based on Bell's Theorem."

23 *In photosynthesis, entanglement*: Vedral, "High-Temperature Macroscopic Entanglement"; Sarovar et al., "Quantum Entanglement in Photosynthetic Light-Harvesting Complexes."

23 *one of the most widely cited articles*: Redner, "Citation Statistics from More Than a Century of Physical Review."

24 *"between philosophy and physics"*: Anton Zeilinger, interview by author, April 1, 2011, New York.

24 *"energy falling in"*: Ramesh Narayan, telephone interview by author, July 19, 2012.

25 *cauldron of swirling gas*: Goss and McGee, "The Discovery of the Radio Source Sagittarius A (Sgr A)."

25 *region is puzzlingly dim*: Broderick, Loeb, and Narayan, "The Event Horizon of Sagittarius A*."

25 *"and vanishing—poof"*: Narayan, telephone interview by author.

25 *"black hole has no surface"*: Ibid.

26 Climbing *magazine*: Johnson, "A Passion for Physical Realms, Minute and Massive."

26 *"intimately relating to nature"*: Steven B. Giddings, e-mail to author, October 27, 2012.

26 *"a big grizzly bear"*: Giddings, interview by author, May 9, 2012. Bits, Branes, Black Holes, Santa Barbara, CA.

26 *"students felt very intimidated"*: Ibid.

27 *just those strings vibrating*: Musser, *The Complete Idiot's Guide to String Theory*.

27 *"swept up in the waves"*: Giddings, interview by author, May 9, 2012.

27 *"maybe this could work"*: Ibid.

27 *"decay to random junk"*: Giddings, e-mail to author, March 30, 2007.

27 *all is lost*: Hawking, "Black Holes and Thermodynamics."

28 *search for escape hatches*: Callan et al., "Evanescent Black Holes."

28 *"Hawking's original picture"*: Giddings, interview by author, May 9, 2012.

28 *we should see parallel failings*: Banks, Susskind, and Peskin, "Difficulties for the Evolution of Pure States into Mixed States."

28 *must be reversible*: Hawking, "Information Loss in Black Holes."
28 *"continuing to beat my head"*: Giddings, interview by author, May 14, 2012. Bits, Branes, Black Holes, Santa Barbara, CA.
28 *much the same conclusion*: Lowe et al., "Black Hole Complementarity Versus Locality."
29 *"I didn't pursue it"*: Giddings, interview by author, May 9, 2012.
29 *that gets people's attention*: Sivers, "How to Start a Movement."
29 *the Hubble Deep Field*: Macchetto and Dickinson, "Galaxies in the Young Universe."
30 *just one photon of their light per minute*: Mario Livio, e-mail to author, November 13, 2012.
30 *first realized in 1969*: Misner, "Mixmaster Universe."
30 *standard textbook of gravity*: Misner, Thorne, and Wheeler, *Gravitation*.
30 *spilling acids on various textiles*: Lightman and Brawer, *Origins: The Lives and Worlds of Modern Cosmologists*, 233.
30 *"The labs were awful"*: Charles W. Misner, telephone interview by author, July 5, 2012.
30 *"a geometrical and physical intuition"*: Ibid.
31 *stretching light waves*: Schmidt, "3C 273 : A Star-Like Object with Large Red-Shift."
31 *the hiss persisted*: Wilson, "The Cosmic Microwave Background Radiation," 475.
31 *13.8 billion years old*: Planck Collaboration, "Planck 2013 Results. XVI. Cosmological Parameters."
31 *breathtakingly unlikely*: Carroll, "In What Sense Is the Early Universe Fine-Tuned?"
32 *"sky is not extremely mottled"*: Misner, telephone interview by author.
32 *Russian theorist Yakov Zel'dovich*: Zel'dovich, "Particle Production in Cosmology."
32 *"at those extreme times"*: Misner, telephone interview by author.
33 *a way to solve the horizon problem*: Guth and Steinhardt, "The Inflationary Universe."
34 *galaxies are not actually moving through space*: Davis and Lineweaver, "Expanding Confusion: Common Misconceptions of Cosmological Horizons and the Superluminal Expansion of the Universe," 5–6.
34 *"a relative velocity"*: Misner, telephone interview by author.
34 *telltale patterns of inflation*: Ade et al., "BICEP2 I: Detection of B-Mode Polarization at Degree Angular Scales."
34 *the finding fizzled*: Planck Collaboration, "Planck Intermediate Results. XXX. The Angular Power Spectrum of Polarized Dust Emission at Intermediate and High Galactic Latitudes."
34 *preternaturally uniform*: Vachaspati and Trodden, "Causality and Cosmic Inflation"; Penrose, *The Road to Reality: A Complete Guide to the Laws of the Universe*, 755–57; Steinhardt, "The Inflation Debate."
34 *shared first place*: Kaplan, "Winners of the Young Researchers Competition in Physics Announced."
35 *"an interesting interplay"*: Fotini Markopoulou, telephone interview by author, November 23, 2012.

35 *"like a planetarium"*: Ibid.

35 *"where Einstein left off"*: Ibid.

35 *"It's a funny thing"*: Markopoulou, interview by author, June 4, 2011, New York.

36 *"the quantum-gravity people"*: Markopoulou, telephone interview by author.

36 *Markopoulou made her name*: Markopoulou and Smolin, "Causal Evolution of Spin Networks"; Markopoulou and Smolin, "Disordered Locality in Loop Quantum Gravity States."

36 *"I had this gut feeling"*: Markopoulou, telephone interview by author.

36 *Several string theorists*: Easther et al., "Constraining Holographic Inflation with WMAP."

36 *"staring us in the face"*: Markopoulou, e-mail to author, May 17, 2012.

36 *most basic function*: Greene, *The Fabric of the Cosmos*, 122.

37 *sheer number of data cables*: Perkins and Malyukov, "Cables: The 'Blood Vessels' of ATLAS."

37 *build a particle detector*: Musser, "How to Build the World's Simplest Particle Detector."

38 *"these Feynman calculations"*: Zvi Bern, e-mail to author, May 16, 2012.

38 *220 different ways*: Mangano and Parke, "Multi-Parton Amplitudes in Gauge Theories," 304.

39 *reduce to a mere four*: Ibid., 326.

39 *"My epiphany about science"*: Bern, interview by author, April 24, 2012, Los Angeles.

39 *jump straight to the final four*: Bern et al., "Fusing Gauge Theory Tree Amplitudes into Loop Amplitudes."

39 *criticized the ayatollahs*: Nima Arkani-Hamed, interview by author, February 23, 2011, Princeton, NJ.

40 *"at a blistering pace"*: Arkani-Hamed, e-mail to author, January 8, 2010.

40 *a comprehensive alternative*: Arkani-Hamed and Trnka, "The Amplituhedron."

40 *the trouble with Feynman diagrams*: Arkani-Hamed, Cachazo, and Kaplan, "What Is the Simplest Quantum Field Theory?"

40 *drawn to Feynman's approach*: Kaiser, *Drawing Theories Apart: The Dispersion of Feynman Diagrams in Postwar Physics*, 368–73.

40 *"You then suffer"*: Arkani-Hamed, telephone interview by author, February 5, 2010.

41 *Particles still obey*: Arkani-Hamed et al., "Local Spacetime Physics from the Grassmannian."

41 *but it fomented revolution*: Kuhn, *The Copernican Revolution: Planetary Astronomy in the Development of Western Thought*, chap. 5.

41 *"When you're a kid"*: Arkani-Hamed, interview by author, July 11, 2011, Princeton, NJ.

42 *parts of the same elephant*: Heller, "Where Physics Meets Metaphysics," 261.

42 *"what entanglement means"*: Arkani-Hamed, interview by author, July 11, 2011.

42 *the glue holding space together*: Giddings, "Black Holes, Quantum Information, and Unitary Evolution"; Van Raamsdonk, "Building Up Spacetime with Quantum Entanglement."

42 *a kind of secret tunnel*: Maldacena and Susskind, "Cool Horizons for Entangled Black Holes."

2. THE ORIGINS OF NONLOCALITY

43 *"end of the rationality"*: Popper, "Bell's Theorem: A Note on Locality," 417.

43 *"incompatible with the very possibility"*: Bohm and Hiley, *The Undivided Universe: An Ontological Interpretation of Quantum Theory*, 157.

43 *"flapdoodle"*: Gell-Mann, *The Quark and the Jaguar: Adventures in the Simple and the Complex*, 172.

43 *the unity in diversity*: Salmon, *Causality and Explanation*, 69–71, 76–78, 85–90.

45 *parallels to the dismay*: McMullin, "The Explanation of Distant Action: Historical Notes," 272.

45 *"a lump of brute matter"*: Frans H. van Lunteren, e-mail to author, August 20, 2012.

45 *"used to missing the point"*: van Lunteren, e-mail to author, August 19, 2012.

45 *"start with a differential equation"*: van Lunteren, e-mail to author, August 18, 2012.

46 *"into the old darkness"*: Hesse, *Forces and Fields: The Concept of Action at a Distance in the History of Physics*, 157.

46 *"hard to swallow"*: van Lunteren, e-mail to author, August 18.

46 *scholars did their utmost*: van Lunteren, "Framing Hypotheses: Conceptions of Gravity in the 18th and 19th Centuries."

46 *recoiled . . . embraced*: Hesse, *Forces and Fields*, 166, 187; Jammer, *Concepts of Force*, 145–46.

46 *twists and turns began*: Cushing, *Quantum Mechanics: Historical Contingency and the Copenhagen Hegemony*, 18.

46 *Plato tells the story*: Taylor, "Parmenides, Zeno, and Socrates."

47 *what drove a person*: Salmon, *Causality*, 6–7.

47 *Poseidon was peeved*: Thucydides, *History of the Peloponnesian War*, 83.

47 *explanations make little distinction*: Hesse, *Forces and Fields*, 34–35.

47 *What rules governed*: Guthrie, *A History of Greek Philosophy*, vol. 1, *The Earlier Presocratics and the Pythagoreans*, 37–45.

47 *floating on a subterranean ocean*: Lloyd, *Early Greek Science: Thales to Aristotle*, 9.

47 *He wasn't so sure*: Guthrie, *A History of Greek Philosophy*, vol. 2, *The Presocratic Tradition from Parmenides to Democritus*, 31–33; Lloyd, *Early Greek Science*, 36–39.

47 *nine such paradoxes*: Huggett, "Zeno's Paradoxes."

47 *"the most self-evident thing"*: Simplicius, *On Aristotle's Physics 6*, 114.

48 *has no innate scale*: Riemann, "On the Hypotheses Which Lie at the Bases of Geometry"; Grünbaum, *Modern Science and Zeno's Paradoxes*, chap. 3.

48 *"One reading of"*: Fay Dowker, interview by author, April 4, 2012. Bits, Branes, Black Holes, Santa Barbara, CA.

48 *"one tiny piece of space/time"*: Feynman, *The Character of Physical Law*, 57.

48 *discrete building blocks*: Guthrie, *History of Greek Philosophy*, vol. 2, 389–92, 455–56; Leucippus and Democritus, *The Atomists: Leucippus and Democritus*, 73–74, 164–71.

48 *sensations we enjoy*: Ibid., 119, 208–209.

49 *the atomists' creation*: Cornford, "The Invention of Space"; Guthrie, *History of Greek Philosophy*, vol. 1, 279; Casey, *The Fate of Place*, 79–84.

49 *"no place or space"*: Lucretius, *The Nature of Things*, 15.

49 *innumerable variety of worlds*: Guthrie, *History*, vol. 2, 404–407.

49 *space separates atoms*: Leucippus and Democritus, *The Atomists*, 184–88.

49 *by making direct contact*: Guthrie, *History of Greek Philosophy*, vol. 2, 418–19, 498; Leucippus and Democritus, *The Atomists* 73, 74, 88.

49 *hardly a phenomenon*: Guthrie, *History of Greek Philosophy*, vol. 2, 388; Leucippus and Democritus, *The Atomists*, 159, 195.

50 *that analogy came centuries later*: Berryman, *The Mechanical Hypothesis in Ancient Greek Natural Philosophy*, 33–39.

50 *Individual atoms are lifeless*: Dijksterhuis, *The Mechanization of the World Picture: Pythagoras to Newton*, 11–12, 495–98.

50 *void of purpose and meaning*: Ibid., 77–78.

50 *to burn his books*: Leucippus and Democritus, *The Atomists*, 56–57.

50 *Pulitzer Prize–winning*: Greenblatt, "The Answer Man."

50 *"rid of harsh taskmasters"*: Lucretius, *The Nature of Things*, 68.

50 *world throbs with life*: Kearney, *Science and Change, 1500–1700*, 23–24, 26–33.

50 *a sophisticated theory*: Newstead, "Aristotle and Modern Mathematical Theories of the Continuum."

50 *the arrangement of atoms*: Hesse, *Forces and Fields*, 62.

51 *Aristotle abhorred a void*: Aristotle, *Physics*, book 4, parts 7–9.

51 *in relation to its neighbors*: Barbour, *Absolute or Relative Motion? The Discovery of Dynamics*, 84–91.

51 *passes the impulse*: Hesse, *Forces and Fields*, 79–80.

51 *Aristotle's contemporaries in China*: Needham and Wang, *Science and Civilisation in China*, vol. 4, part 1, 6–8.

51 *Aristotle's way of thinking*: Lloyd, *Early Greek Science*, 103–9; Rovelli, "Aristotle's Physics: A Physicist's Look."

51 *corroborate or disprove*: Lloyd, *Early Greek Science*, 139–42.

51 *making the universe comprehensible*: Hesse, *Forces and Fields*, 72.

51 *purely by contact*: Ibid., 67–73.

51 *"bodily change of place"*: Aristotle, *Physics*, book 7, part 1, 242b59.

51 *Aristotle took pains*: Jammer, *Concepts of Space: The History of Theories of Space in Physics*, 17–22.

51 *"a goat-stag or a sphinx"*: Aristotle, *Physics*, book 4, part 1, 208a27.

51 *many strange rocks*: Hesse, *Forces and Fields*, 57–58.

51 *known as Magnesia*: Melfos, Helly, and Voudouris, "The Ancient Greek Names 'Magnesia' and 'Magnetes' and Their Origin from the Magnetite Occurrences at the Mavrovouni Mountain of Thessaly, Central Greece. A Mineralogical–Geochemical Approach."

51 *Thales also marveled at amber*: Guthrie, *History of Greek Philosophy*, vol. 1, 66.

51 *Greek word for amber*: Gilbert, *On the Magnet*, 46–47.

51 *Chinese scholars discovered*: Needham and Wang, *Science and Civilisation in China*, vol. 4, part 1, 230, 232.

52 *vapors that displaced the air*: Guthrie, *History of Greek Philosophy*, vol. 2, 426.

52 *ignoring it*: Hesse, *Forces and Fields*, 57.

52 *rotating crystalline spheres*: Kuhn, *The Copernican Revolution*, 79–80; Jammer, *Concepts of Force*, 41–42.

52 *swept toward the middle*: Guthrie, *History of Greek Philosophy*, vol. 2, 400–413; Leucippus and Democritus, *The Atomists*, 94–95, 179–84.

53 *"nothing is distant"*: Saint Thomas Aquinas, *The Summa Theologica of Saint Thomas Aquinas*, book 1, query 8, article 1, reply to objection 3.

53 *The more scholars mulled*: Kuhn, *The Copernican Revolution*, 101–104, 115–17.

53 *a lodestone attracts*: Hesse, *Forces and Fields*, 87–90.

53 *an act of "coition"*: Gilbert, *On the Magnet*, 1, 208.

54 *evidence for atoms*: Kuhn, *The Copernican Revolution*, 89–90, 231–37.

54 *"all the phenomena of nature"*: Descartes, *The Philosophical Writings of Descartes*, vol. 3, *The Correspondence*, 7.

54 *as comprehensive as Aristotle's*: Dijksterhuis, *The Mechanization of the World Picture*, 408.

54 *served as the manifesto*: Kearney, *Science and Change*, 151–60.

54 *continuities are clear*: Dijksterhuis, *The Mechanization of the World Picture*, 417; Garber, *Descartes' Metaphysical Physics*, 119.

54 *purely geometric figure*: Descartes, *Principles of Philosophy*, 40–41.

54 *freely in straight lines*: Suppes, "Descartes and the Problem of Action at a Distance," 149–50.

54 *"require no proof"*: Descartes, *Principles of Philosophy*, 69.

55 *inspired modern theories*: Westfall, *The Construction of Modern Science: Mechanisms and Mechanics*, 34–36.

55 *tiny screws or hooks*: Hesse, *Forces and Fields*, 106–107.

55 *alchemy lab in a garden shed*: Westfall, "Newton and Alchemy," 318–19.

56 *Feynman had been fascinated*: Feynman, *"Surely You're Joking, Mr. Feynman!": Adventures of a Curious Character*, 332–36.

56 *a product of magic*: Kearney, *Science and Change*, 22–25, 48; Westfall, "Newton and the Hermetic Tradition," 195; Henry, *The Scientific Revolution and the Origins of Modern Science*, 55.

56 *not enough heart*: Goodrick-Clarke, *The Western Esoteric Traditions: A Historical Introduction*, chap. 1.

56 *the oneness of nature*: Stamatellos, *Plotinus and the Presocratics: A Philosophical Study of Presocratic Influences in Plotinus' Enneads*, 26–29, 150–54.

56 *They remain influential*: Hanegraaff, "The New Age Movement and the Esoteric Tradition."

56 *charms and potions*: Copenhaver, "Natural Magic, Hermetism, and Occultism in Early Modern Science," 270–81; Hesse, *Forces and Fields*, 31–32; Goodrick-Clarke, *The Western Esoteric Traditions*, 8, 56.

56 *"their causes lie hid"*: Agrippa von Nettesheim, *Three Books of Occult Philosophy*, 32.

57 *chocolate frog trading card*: Rowling, *Harry Potter and the Sorcerer's Stone*, 102.

57 *an essential part*: Sack, "Magic and Space."

57 *sympathies and antipathies*: Jammer, *Concepts of Force*, 42–47.

57 *"like produces like"*: Frazer, *The Golden Bough: A Study in Magic and Religion; Part 1: The Magic Art and and the Evolution of Kings*, vol. 1, 52.

57 *alchemy, astrology, and numerology*: Hesse, *Forces and Fields*, 74–77, 93–97.

57 *Western culture has seesawed*: Brush, "The Chimerical Cat: Philosophy of Quantum Mechanics in Historical Perspective," 403–10.

57 *traces of it in the Bohr-Einstein debates*: Forman, "Weimar Culture, Causality, and Quantum Theory, 1918–1927: Adaptation by German Physicists and Mathematicians to a Hostile Intellectual Environment," 111–12; Brush, "The Chimerical Cat," 410–18.

58 *Agrippa latched on to them*: Kearney, *Science and Change*, 40–41.

58 *To probe its mysteries*: Henry, *The Scientific Revolution*, chap. 4.

58 *to manipulate nature*: Yates, *Giordano Bruno and the Hermetic Tradition*, 155–56.

58 *Pico della Mirandola*: Goodrick-Clarke, *The Western Esoteric Traditions*, 45.

58 *For that noble sentiment*: Yates, *Giordano Bruno*, 112.

58 *Hamlet soliloquy*: Caldiero, "The Source of Hamlet's 'What a Piece of Work Is a Man!'"

58 *green lizard urine*: Agrippa von Nettesheim, *Three Books of Occult Philosophy*, 150.

58 *laying the foundations*: Kearney, *Science and Change*, 116–18, 130–32.

58 *"The aim of magic"*: Rossi, *Francis Bacon: From Magic to Science*, 22.

59 *magical notion of sympathies*: Jammer, *Concepts of Force*, 72.

59 *crises of self-doubt*: Koestler, *The Watershed: A Biography of Johannes Kepler*, 123–24.

59 *"here I blundered"*: Ibid., 62.

59 *his mystical inspiration*: Kearney, *Science and Change*, 138.

59 *Kepler cast horoscopes*: Koestler, *The Watershed*, 39–42.

59 *if the moon were watery*: Kepler, *New Astronomy*, 56–57; McMullin, "The Origins of the Field Concept in Physics," 18–19.

59 *fine-tune planet orbits*: Kuhn, *The Copernican Revolution*, 246.

59 *purists were unsympathetic*: McMullin, "Origins of the Field Concept," 20.

59 *"the moon's dominion"*: Galilei, *Dialogue Concerning the Two Chief World Systems, Ptolemaic and Copernican*, 462.

59 *fringes of mainstream philosophy*: Koestler, *The Watershed*, 163–65.

59 *"the great amphibian"*: Kearney, *Science and Change*, 196.

59 *as flaming atheists*: Henry, "Occult Qualities and the Experimental Philosophy: Active Principles in Pre-Newtonian Matter Theory," 352–53.

60 *alchemy, Neoplatonism, and Kabbalah*: Ibid., 352–53, 357–58; Westfall, *The Life of Isaac Newton*, 117–19.

60 *bore witness*: Dobbs, "Newton's Alchemy and His Theory of Matter," 526–27.

60 *"intellectually sleep around"*: Albion Lawrence, interview by author, April 12, 2012. Bits, Branes, Black Holes, Santa Barbara, CA.

60 *knew perfectly well*: Kearney, *Science and Change*, 194–96.

60 *congratulating him*: Newton, *Isaac Newton: Philosophical Writings*, 106–107.

61 *fifteen thousand letters*: Leibniz, *Philosophical Papers and Letters*, vol. 1, 549n21.

61 *"how extraordinarily distracted"*: Mates, *Philosophy of Leibniz*, 27.

61 *five rounds of letters*: Westfall, *Life of Isaac Newton*, 294; Leibniz, *Philosophical Papers and Letters*, vol. 2, 675.

61 *what would qualify as a satisfying resolution*: Kuhn, *The Structure of Scientific Revolutions*, 109.

61 *a mechanical explanation*: Mischel, "Pragmatic Aspects of Explanation."

61 *unexplainable*: Hutchison, "What Happened to Occult Qualities in the Scientific Revolution?," 253.

61 *"a chimerical thing"*: Leibniz, *Philosophical Papers and Letters*, vol. 2, 716.

62 *"I frame no hypotheses"*: Newton, *The Mathematical Principles of Natural Philosophy*, vol. 1, 314.

62 *that's good enough*: Henry, "Occult Qualities and the Experimental Philosophy," 358–59, 362–63.

62 *two separate functions*: Janiak, *Newton as Philosopher*, 15–25, 53–65.

62 *a compelling picture*: Jammer, *The Philosophy of Quantum Mechanics: The Interpretations of Quantum Mechanics in Historical Perspective*, 10–11.

62 *An insistence on explanation*: Hutchison, "What Happened to Occult Qualities in the Scientific Revolution?," 251.

62 *"restrain the intemperate desire"*: Hume, *A Treatise of Human Nature*, book 1, *Of the Understanding*, 13.

62 *known as instrumentalism*: Popper, "Three Views Concerning Human Knowledge."

63 *reassure their colleagues*: Kuhn, *The Copernican Revolution*, 229–30; Beller, *Quantum Dialogue: The Making of a Revolution*, 176–77.

63 *the creative spark of science*: Lightman, "Magic on the Mind."

63 *no sharp boundary*: Lange, *An Introduction to the Philosophy of Physics: Locality, Fields, Energy, and Mass*, 249–50.

63 *three broad categories*: Newton, *Mathematical Principles*, vol. 1, 174; Hesse, *Forces and Fields*, 148–53; McMullin, "The Explanation of Distant Action," 293–301.

64 *particles impart a force*: Newton, *Mathematical Principles*, vol. 2, 313.

64 *violent religious fanatics*: van Lunteren, "Nicolas Fatio de Duillier on the Mechanical Cause of Universal Gravitation."

64 *"immaterial"*: Newton, *Philosophical Writings*, 102.

64 *"incorporeal"*: Newton, *Mathematical Principles*, vol. 1, 174.

64 *"intangible"*: Leibniz, *Philosophical Papers and Letters: A Selection*, vol. 2, 696.

64 *penetrate the interior of planets*: Janiak, *Newton as Philosopher*, 76–79; Kochiras, "Gravity and Newton's Substance Counting Problem."

64 *inspired the concept*: Gale, "Leibniz and Some Aspects of Field Dynamics"; Friedman, Introduction to Kant's *Metaphysical Foundations of Natural Science*, ix–x.

65 *God's omnipresence*: Newton, *Philosophical Writings*, 22, 25–27.

65 *God already exists*: Kochiras, "Gravity and Newton's Substance Counting Problem," 270–72; Janiak, *Newton as Philosopher*, 37–40.

65 *monads gave rise*: Gale, "Leibniz and Some Aspects of Field Dynamics," 39–40; Slowik, "The 'Properties' of Leibnizian Space: Whither Relationism?," 123–24, 128–29.

65 *Neither man thought*: Newton, *Philosophical Writings*, 21–22; Leibniz, *Leibniz: New Essays on Human Understanding*, book 2, chap. 13, paragraph 17.

65 *helped Newton revise*: Janiak, *Newton as Philosopher*, 90–94, 168–72.

65 *referring to atheism*: Newton, *Philosophical Writings*, 102; Henry, "'Pray Do Not Ascribe That Notion to Me': God and Newton's Gravity."

65 *invoked nonlocal forces*: Hesse, *Forces and Fields*, 153–56, 180–88; Henry, "Gravity and De Gravitatione: The Development of Newton's Ideas on Action at a Distance," 20–23.

65 *alienate mechanical purists*: Westfall, *Life of Isaac Newton*, 187–88.

65 *perfectly reasonable*: van Lunteren, "Framing Hypotheses," 68–90; Hesse, *Forces and Fields*, 155–56, 166, 187.

65 *fluids for magnetism*: Williams, *The Origins of Field Theory*, 17–27; Hesse, *Forces and Fields*, 182–83.

65 *did a backflip*: Ibid., 166.

66 *supposedly simple collision*: Jammer, *Concepts of Force*, 211.

66 *proponents of locality*: Leibniz, *Philosophical Papers and Letters*, vol. 2, 446; Leucippus and Democritus, *The Atomists*, 84, 186–87, 192–93.

66 *an instant U-turn*: Hesse, *Forces and Fields*, 163–66.

66 *German überphilosopher played*: Kuehn, *Kant: A Biography*, 64.

66 *a continuum of forces*: Friedman, "Introduction" pp. xvi–xix; Kant, *Metaphysical Foundations of Natural Science*, 34, 50–55; Williams, *Origins of Field Theory*, 37–43.

66 *never actually come into contact*: Hesse, *Forces and Fields*, 163–66.

66 *reduced local forces*: van Lunteren, "Framing Hypotheses," 126.

67 *"reduce uncommon unintelligibilities"*: Mach, *History and Root of the Principle of the Conservation of Energy*, 55–56.

67 *"gravitation no longer disturbs"*: Mach, *History and Root of the Principle of the Conservation of Energy*, 56.

67 *frogs' legs twitch*: Kipnis, "Luigi Galvani and the Debate on Animal Electricity, 1791–1800," 114.

67 *a wet piece of cardboard*: Ibid., 135.

67 *an amazing new plaything*: Berkson, *Fields of Force: The Development of a World View from Faraday to Einstein*, 30.

68 *rebellion against mechanistic thinking*: Williams, *Origins of Field Theory*, 31, 43.

68 *revivals of magical thinking*: Goodrick-Clarke, *The Western Esoteric Traditions*, 180.

68 *Adherents were fascinated*: Safranski and Osers, *Schopenhauer and the Wild Years of Philosophy*, 200–201.

68 *nature's diverse forces*: Morus, *When Physics Became King*, 54–63.

68 *built his first battery*: Stauffer, "Speculation and Experiment in the Background of Oersted's Discovery of Electromagnetism," 40.

68 *no magnetic effects*: Ibid., 43.

68 *a wire connected to a battery*: Ibid., 46.

68 *blind to other objects*: Berkson, *Fields of Force* , 21–22.

69 *the swirling motions*: Maxwell, *The Scientific Papers of James Clerk Maxwell*, vol. 2, 317.

69 *accepted Newton's atomistic explanation*: Cantor, *Optics After Newton: Theories of Light in Britain and Ireland, 1704–1840*, 29–31, 86–90, 204.

69 *inspiration from the flow*: Ibid., 129–30.

69 *among medieval scholastics*: Hesse, *Forces and Fields*, 81.

69 *use a laser pointer*: Prutchi and Prutchi, *Exploring Quantum Physics Through Hands-On Projects*, 4–6.

69 *Under the emperor*: Morus, *When Physics Became King*, 23, 26–32.

69 *apt to misinterpret*: Cantor, *Optics After Newton: Theories of Light in Britain and Ireland, 1704–1840*, 142–44.

70 *a bookbinder's apprentice*: Williams, *Michael Faraday: A Biography*, 8, 14.

70 *borrowed a shilling*: Ibid., 15, 22.

71 *their vision of unity*: Williams, *Origins of Field Theory*, 68–69; Morus, *When Physics Became King*, 91–97.

71 *The word "physicist"*: Morus, 6–7, 53.

71 *a rebranding strategy*: Ibid.

71 *never learned math*: Williams, *Origins of Field Theory*, 67.

71 *nature is local*: Berkson, *Fields of Force*, 39–49.

71 *the lines of force*: Williams, *Origins of Field Theory*, 76.

71 *ordinary substance made of*: Doran, "Origins and Consolidation of Field Theory in Nineteenth-Century Britain: From the Mechanical to the Electromagnetic View of Nature," 164–65; Hesse, *Forces and Fields*, 199–200.

72 *an immaterial medium*: Faraday, *Experimental Researches in Electricity*, vol. 2, 284–93, vol. 3, 447–52; Doran, "Origins and Consolidation of Field Theory," 166–78.

72 *introduced the term*: Faraday, *Experimental Researches in Electricity*, vol. 3, 30.

72 *failed to catch on*: Williams, *Origins of Field Theory*, 112–13, 117.

72 *at points in space*: Maxwell, *Scientific Papers*, vol. 1, 160, 205.

73 *could equally well*: Maxwell, *A Treatise on Electricity and Magnetism*, vol. 1, x.

73 *describe a hypothetical*: Hesse, *Forces and Fields*, 196–98.

73 *disturbances take time*: Berkson, *Fields of Force*, 231–40.

73 *A time lag seems odd*: Faraday, *Experimental Researches in Electricity*, vol. 3, 409, 412.

73 *an electromagnetic wave*: Maxwell, *Scientific Papers*, vol. 1, 535, 579–80.

73 *essence of real things*: Ibid., 564; Lange, *Introduction to the Philosophy of Physics*, 120–36.

73 *none goes missing*: Faraday, *Experimental Researches in Chemistry and Physics*, 443–63.

75 *"most pernicious heresy"*: Tait, *Properties of Matter*, 6; Lodge, *Modern Views of Electricity*, 331.

75 *no place in Maxwell's theory*: van Lunteren, "Gravitational and Nineteenth-Century Physical Worldviews."

75 *gravity always pulls*: Maxwell, *Scientific Papers*, vol. 1, 570–71.

75 *zipped across space*: Hesse, *Forces and Fields*, 195, 225.

75 *single out a certain speed*: Lange, *Introduction to the Philosophy of Physics*, 210–12.

76 *like they're standing still*: Einstein, "Autobiographical Notes," in Schilpp, *Albert Einstein, Philosopher-Scientist*, 53.

76 *were equally contradictory*: Berkson, *Fields of Force*, 261–67.

76 *like a bear in a cage*: de Haas-Lorentz, *H. A. Lorentz: Impressions of His Life and Work*, 41.

76 *perhaps even of gravity*: McCormmach, "H. A. Lorentz and the Electromagnetic View of Nature."

76 *relative to the apparatus*: Berkson, *Fields of Force*, 274–75, 313–15.

76 *awkward consequences*: Einstein, "Physics and Reality," 364–65; Berkson, *Fields of Force*, 271; Frisch, "Inconsistency in Classical Electrodynamics."

77 *as though it were psychic*: Dirac, "Classical Theory of Radiating Electrons," 159–60; Jackson, *Classical Electrodynamics*, 786–98.

77 *used to send messages*: Frisch, "Non-Locality in Classical Electrodynamics," 4–7.

77 *popping off like firecrackers*: Dirac, "Classical Theory of Radiating Electrons," 149.

77 *an infinite capacity*: Rayleigh, "The Dynamical Theory of Gases and of Radiation."

77 *physicists came to question*: Ritz, "Recherches critiques sur l'électrodynamique générale"; Wheeler and Feynman, "Classical Electrodynamics in Terms of Direct Interparticle Action."

3. EINSTEIN'S LOCALITY

79 *all the fun stuff*: Isaacson, *Einstein: His Life and Universe*, 32–37.

79 *great works of philosophy*: Howard, "Albert Einstein as a Philosopher of Science."

79 *rejected job applications*: Isaacson, *Einstein*, 54–61.

79 *his first scientific papers*: McCormmach, "Einstein, Lorentz, and the Electron Theory," 43–44.

79 *tweaked those equations*: Norton, "Einstein's Investigations of Galilean Covariant Electrodynamics Prior to 1905."

80 *Einstein's egalitarian instincts*: Zahar, "Why Did Einstein's Programme Supersede Lorentz's? (II)," 232–33.

80 *quintessential eureka moment*: Abiko, "Einstein's Kyoto Address: 'How I Created the Theory of Relativity,'" 14.

80 *addition or subtraction*: Newton, *The Mathematical Principles of Natural Philosophy*, vol. 1, 21–22.

80 *assumption of instantaneous communication*: Ehlers, "The Nature and Structure of Spacetime," 73–74.

80 *confront their own ignorance*: Fernbach et al., "Political Extremism Is Supported by an Illusion of Understanding."

80 *exchange some kind of signal*: Einstein, "On the Electrodynamics of Moving Bodies," 126–27.

81 *alternative to the Newtonian velocity-addition rule*: Einstein and Penrose, *Einstein's Miraculous Year: Five Papers That Changed the Face of Physics*, 142.

81 *to countenance nonlocality*: McCormmach, "Einstein, Lorentz, and the Electron Theory," 67.

81 *pointed out some repercussions*: Minkowski, "Raum Und Zeit."

81 *an objective fact*: Nozick, *Invariances: The Structure of the Objective World*, 76–77.

82 *a property of space*: Carroll, *Spacetime and Geometry: An Introduction to General Relativity*, 2, 48–50, 178–79.

83 *theory doesn't forbid*: Liberati, Sonego, and Visser, "Faster-Than-C Signals, Spe-

cial Relativity, and Causality"; Maudlin, *Quantum Non-Locality and Relativity*, 99–116; Hesse, *Forces and Fields*, 245.

83 *the end of a rainbow*: Maudlin, *Quantum Non-Locality and Relativity*, 70.

83 *a factor of 10*: Lincoln, "Proving Special Relativity: Episode 2."

83 *Reaching light speed*: Einstein, "On the Electrodynamics of Moving Bodies," 158.

84 *muck up sequences*: Hesse, *Forces and Fields*, 236, 283–88, 305.

84 *as early as 1907*: Einstein, "Über die vom Relativitätsprinzip geforderte Trägheit der Energie," 381–82; Schwartz, "Einstein's Comprehensive 1907 Essay on Relativity, Part 1," 516; Bell, *Speakable and Unspeakable in Quantum Mechanics*, 235–36.

84 *"telegraph into the past"*: Langevin, "L'évolution de l'espace et du temps," 44; Miller, *Albert Einstein's Special Theory of Relativity: Emergence (1905) and Early Interpretation (1905–1911)*, 223.

85 *creating a causal loop*: Mitchell, "The Clock That Went Backward."

85 *preventing your own birth*: Schachner, "Ancestral Voices."

85 *life imitating art*: Gödel, "A Remark About the Relationship Between Relativity Theory and Idealistic Philosophy," 560–61; Black, "Why Cannot an Effect Precede Its Cause?," 54–55.

85 *acquainted with anarchy*: Borisov and Kudryashov, "Paul Painlevé and His Contribution to Science."

85 *at infinite velocity*: Saari and Xia, "Off to Infinity in Finite Time."

85 *a space invader*: Earman, *A Primer on Determinism*, 46–47.

86 *allow pulses to ripple*: Ibid., 40–42.

86 *your laundry basket*: Ibid., 55–61.

86 *"the potential to be garbage"*: Giddings, interview by author, October 11, 2010, Santa Barbara, CA.

86 *Even Niels Bohr*: Bohr, "Space and Time in Nuclear Physics," 212, 218.

86 *every point of a field*: Einstein, "Quanten-Mechanik Und Wirklichkeit," 321–22; Howard, "Holism, Separability and the Metaphysical Implications of the Bell Experiments," 232–40.

86 *it was a supposition*: Einstein, "Autobiographical Notes," 13; Fine, *The Shaky Game: Einstein, Realism, and the Quantum Theory*, chap. 6.

87 *"conceived in sin"*: Arthur Fine, e-mail to author, June 24, 2011.

87 *admitted they didn't know*: Stachel, *Einstein from 'B' to 'Z,'* 378–79.

87 *the theory's father*: Stone, "Einstein and the Quantum: The Quest of the Valiant Swabian," 281–82.

87 *practically the only one*: Stuewer, "The Experimental Challenge of Light Quanta," 146–47.

87 *Einstein settled the issue*: Einstein, "On a Heuristic Point of View Concerning the Production and Transformation of Light"; Einstein, "Über die Entwicklung unserer Anschauungen über das Wesen und die Konstitution der Strahlung."

88 *nonlocality seems unavoidable*: Hesse, *Forces and Fields*, 265.

88 *atoms absorb wave energy in discrete bites*: Stuewer, "The Experimental Challenge of Light Quanta," 147–48.

88 *saw very early on*: Bacciagaluppi and Valentini, *Quantum Theory at the Crossroads: Reconsidering the 1927 Solvay Conference*, 178–81.

88 *the "bubble paradox"*: Cramer, "The Quantum Handshake."

88 *Einstein's initial instinct*: McCormmach, "Einstein, Lorentz, and the Electron Theory," 56–57.

88 *particles acting independently*: Howard, "'Nicht Sein Kann Was Nicht Sein Darf,' or the Prehistory of EPR, 1909–1935: Einstein's Early Worries About the Quantum Mechanics of Composite Systems," 69–78.

89 *"guiding field"*: Ibid., 72–73, 75–76.

89 *Bohr toyed with a version*: Howard, "Revisiting the Einstein-Bohr Dialogue," 67–69.

90 *"ordinary space-time description"*: Bohr, *The Emergence of Quantum Mechanics (Mainly 1924–1926)*, vol. 5, 79.

90 *to take a break*: Isaacson, *Einstein: His Life and Universe*, 157.

90 *"a kind of 'foamy crest'"*: Schrödinger, "On Einstein's Gas Theory."

91 *a curious mathematical abstraction*: Schrödinger, "Quantisation as a Problem of Proper Values (Part IV)," 120.

91 *admitted he didn't know*: Tollaksen et al., "Quantum Interference Experiments, Modular Variables and Weak Measurements," 5n6.

91 *distant regions of space*: Aharonov and Rohrlich, *Quantum Paradoxes*, 61–75.

91 *two sides to the debate*: Beller, *Quantum Dialogue: The Making of a Revolution*, 143, 187–89.

92 *spits out a photon*: Einstein, "On the Quantum Theory of Radiation," 76.

92 *"literally, a miracle"*: Earman, "Locality, Nonlocality, and Action at a Distance: A Skeptical Review of Some Philosophical Dogmas," 475.

92 *traced this cultural mood*: Forman, "Weimar Culture, Causality, and Quantum Theory, 1918–1927: Adaptation by German Physicists and Mathematicians to a Hostile Intellectual Environment," 111–12; Brush, "The Chimerical Cat: Philosophy of Quantum Mechanics in Historical Perspective," 410–18.

92 *God does not throw dice*: Born and Einstein, *The Born-Einstein Letters 1916–1955: Friendship, Politics and Physics in Uncertain Times*, x, 88, 146.

92 *never objected to randomness*: von Plato, *Creating Modern Probability: Its Mathematics, Physics and Philosophy in Historical Perspective*, 114–23.

92 *give up the search*: Popper, *Quantum Theory and the Schism in Physics*, xvii, 7–10.

93 *not half and half*: Stachel, *Einstein from 'B' to 'Z*,' 410–14.

93 *would entail nonlocality*: Howard, "The Shaky Game," 130–35; Bell, *Speakable and Unspeakable in Quantum Mechanics*, 143.

93 *this dilemma*: Einstein, "Autobiographical Notes," 682; Redhead, *Incompleteness, Nonlocality, and Realism: A Prolegomenon to the Philosophy of Quantum Mechanics*, 76.

93 *Belgian chemicals magnate*: Bacciagaluppi and Valentini, *Quantum Theory at the Crossroads*, 3, 18–19.

93 *twenty-eight dapper men and one elegant woman*: Ibid., 257.

93 *orchestrate the collapse*: Ibid., 440–42.

94 *"to my mind a contradiction"*: Ibid., 441.

94 *name of "realism"*: Laudisa, "Non-Local Realistic Theories and the Scope of the Bell Theorem."

94 *a "hidden variable"*: Bohm, "A Suggested Interpretation of the Quantum Theory in Terms of 'Hidden' Variables, I," 168.

94 *Einstein had been trying*: Belousek, "Einstein's 1927 Unpublished Hidden-Variable Theory: Its Background, Context and Significance."

94 *de Broglie presented*: Bacciagaluppi and Valentini, *Quantum Theory at the Crossroads*, 67–76, 341–71.

94 *"Mr. de Broglie is right"*: Ibid., 441.

94 *Bohr implicitly accepted*: Howard, "Revisiting the Einstein-Bohr Dialogue," 59; Landsman, "When Champions Meet: Rethinking the Bohr-Einstein Debate," 233–34.

94 *"some mathematical methods"*: Bacciagaluppi and Valentini, *Quantum Theory at the Crossroads*, 442.

94 *Heisenberg's uncertainty principle*: Bohr, "Discussion with Einstein on Epistemological Problems in Atomic Physics," 213–18; Howard, " 'Nicht Sein Kann Was Nicht Sein Darf,' " 93–97.

95 *"like a jack-in-the-box"*: Bohr, *Foundations of Quantum Physics I (1926–1932)*, 38.

95 *"so very repugnant"*: Einstein, Letter to Paul Epstein (EA 10-583); Howard, " 'Nicht Sein Kann Was Nicht Sein Darf,' " 102.

96 *either nonlocal or incomplete*: Howard, " 'Nicht Sein Kann Was Nicht Sein Darf,' " 98–105.

96 *this extra functionality is secondary*: Fine, *The Shaky Game*, 37–38; Maudlin, *Quantum Non-Locality and Relativity*, 139–40.

96 *"That which really exists"*: Born and Einstein, *The Born-Einstein Letters*, 162–63.

96 *A predecessor of mine*: Musser, "Forces of the World, Unite!"

96 *stayed up half the night*: Bohr, "Discussion with Einstein on Epistemological Problems in Atomic Physics," 224–28; Isaacson, *Einstein: His Life and Universe*, 347–49.

97 *never to return*: Isaacson, *Einstein: His Life and Universe*, 405–409.

97 *three letters a day*: Gilder, *The Age of Entanglement*, chap. 16.

97 *coined a name*: Schrödinger, "Discussion of Probability Relations Between Separated Systems."

97 *famous morbid scenario*: Trimmer, "The Present Situation in Quantum Mechanics: A Translation of Schrödinger's 'Cat Paradox' Paper."

97 *blamed his coauthors*: Einstein, Letter to Erwin Schrödinger (EA 22-47).

97 *attempted a takedown*: Howard, "Einstein on Locality and Separability."

97 *his own version*: Einstein, "Physics and Reality," 376.

97 *came out with a rebuttal*: Fine, *The Shaky Game*, 191–92; Beller, *Quantum Dialogue*, 153, 277.

98 *"Bohr won the debate"*: Milburn, *The Feynman Processor: Quantum Entanglement and the Computing Revolution*, 47.

98 *the dominant interpretation*: Schlosshauer, Kofler, and Zeilinger, "A Snapshot of Foundational Attitudes Toward Quantum Mechanics."

98 *a loner*: Isaacson, *Einstein: His Life and Universe*, 273–75.

98 *Bohr was a father figure*: Beller, *Quantum Dialogue*, 177, 244, 270.

98 *He and his acolytes*: Ibid., 10–11.

98 *played down Einstein's*: Klein, "Einstein and the Wave-Particle Duality," 3–4.

98 *"misunderstanding"*: Rosenfeld, "Niels Bohr in the Thirties: Consolidation and Extension of the Conception of Complementarity," 128.

98 *"intelligent and promising"*: Bohr, *Foundations of Quantum Physics II (1933–1958)*, 251.

98 *counterfactual scenarios*: Cushing, *Quantum Mechanics*, chap. 10.

4. THE GREAT DEBATE

99 *"a huge bloody argument"*: Maudlin, interview by author.

100 *an intellectual "morass"*: Name withheld, interview by author, November 15, 2010, Utrecht, Netherlands.

100 *"high opinion of themselves"*: Name withheld, interview by author, December 7, 2011, Singapore.

100 *"like a battering ram"*: Name withheld, interview by author, June 7, 2011, New York.

100 *"unassailably appallingly wrong"*: Name withheld, e-mail to author, May 15, 2008.

100 *"war of all against all"*: Fine, e-mail to author, June 24, 2011.

101 *"code for emotional differences"*: Ibid.

101 *Einstein's original paper*: Einstein, Podolsky, and Rosen, "Can Quantum-Mechanical Description of Physical Reality Be Considered Complete?"

101 *John Bell's follow-up*: Bell, "On the Einstein-Podolsky-Rosen Paradox."

101 *"smothered by the formalism"*: Einstein, Letter to Erwin Schrödinger (EA 22–47); Fine, *The Shaky Game: Einstein, Realism, and the Quantum Theory*, 35.

101 *two distinct steps*: Maudlin, *Quantum Non-Locality and Relativity: Metaphysical Intimations of Modern Physics*, 144; Laudisa, "Non-Local Realistic Theories and the Scope of the Bell Theorem," 1122.

102 *stop the presses*: Belousek, "Einstein's 1927 Unpublished Hidden-Variable Theory: Its Background, Context and Significance."

103 *created a video*: Musser, "George and John's Excellent Adventures in Quantum Entanglement."

103 *about 85 percent of the time*: Gisin, "Can Relativity Be Considered Complete? From Newtonian Nonlocality to Quantum Nonlocality and Beyond."

104 *any number of particles*: Greenberger et al., "Bell's Theorem Without Inequalities."

104 *Bohm reimagined it*: Bohm, "A Suggested Interpretation of the Quantum Theory in Terms of 'Hidden' Variables. II," 186–87.

104 *not just of creating dainty patterns*: Maudlin, *Quantum Non-Locality and Relativity*, 119.

104 *it might be possible*: Valentini, "Beyond the Quantum."

104 *rather than attempting to conceal it*: Maudlin, *Quantum Non-Locality and Relativity*, 120–21.

105 *the most stubborn*: Mitroff, "Norms and Counter-Norms in a Select Group of the Apollo Moon Scientists: A Case Study of the Ambivalence of Scientists."

105 *supposedly airtight arguments*: Bauer, *Scientific Literacy and the Myth of the Scientific Method*, 73–78.

105 *ways to stay in sync*: Griffiths, "Quantum Locality"; Unruh, "Minkowski Space-Time and Quantum Mechanics"; Weatherall, "The Scope and Generality of Bell's Theorem."

105 *can't draw any conclusions*: Smerlak and Rovelli, "Relational EPR."

106 *the physicist Nick Herbert*: Kaiser, *How the Hippies Saved Physics: Science, Counterculture, and the Quantum Revival*, chap. 9.

107 *the supposed link*: de Muynck, "Can We Escape from Bell's Conclusion That Quantum Mechanics Describes a Non-Local Reality?" 316–17.

107 *"can't send a signal"*: Giddings, interview by author, October 11, 2010.

107 *"in standard quantum mechanics"*: Maudlin, e-mail to author, October 18, 2012.

108 *the name of "superdeterminism"*: Davies and Brown, *The Ghost in the Atom: A Discussion of the Mysteries of Quantum Physics*, 47.

108 *choices can be entirely open*: Ismael, "Decision and the Open Future"; List, "Free Will, Determinism, and the Possibility of Doing Otherwise."

108 *debate the question*: Grim, "Free Will in Context: A Contemporary Philosophical Perspective."

108 *set up in advance*: Price, *Time's Arrow and Archimedes' Point: New Directions for the Physics of Time*, 234–40; Bell, *Speakable and Unspeakable in Quantum Mechanics*, 100–104, 154.

109 *crazy-sounding ideas*: 't Hooft, "Discreteness and Determinism in Superstrings," 14, 21–22.

109 *"impossible to formulate"*: Gerard 't Hooft, e-mail to author, December 31, 2009.

109 *"looks like a conspiracy"*: 't Hooft, "The Future of Quantum Mechanics."

109 *that slight restriction*: Barrett and Gisin, "How Much Measurement Independence Is Needed to Demonstrate Nonlocality?"; Hall, "Relaxed Bell Inequalities and Kochen-Specker Theorems."

109 *"Constraints are common"*: Fine, e-mail to author, June 24, 2011.

109 *have yet to occur*: Price, *Time's Arrow and Archimedes' Point*, chaps. 5, 8.

110 *particles are precognitive*: Wheeler and Feynman, "Classical Electrodynamics in Terms of Direct Interparticle Action."

110 *"Reverse causation can"*: Maudlin, e-mail to author, March 31, 2011.

110 *we don't routinely see*: Price, *Time's Arrow and Archimedes' Point*, chap. 7.

110 *not to open a portal into the past*: Ibid., 128–29, 173–74, 243–44, 247–48, 250.

111 *sees the coin turn up heads*: Wallace, *The Emergent Multiverse: Quantum Theory According to the Everett Interpretation*, 36–38.

112 *near-duplicates of you*: DeWitt, "Quantum Mechanics and Reality."

112 *minus the cosmological prolificacy*: Gell-Mann and Hartle, "Quantum Mechanics in the Light of Quantum Cosmology," 340; Smerlak and Rovelli, "Relational EPR"; Griffiths, "EPR, Bell, and Quantum Locality"; Maudlin, *Quantum Non-Locality and Relativity*, 216–20.

112 *no need for nonlocal influences*: Bacciagaluppi, "Remarks on Space-Time and Locality in Everett's Interpretation."

113 *"looking out for hours"*: Zeilinger, interview by author.

113 *"the fundamental questions"*: Ibid.

113 *"Something was missing"*: Ibid.

114 *he rather likes it*: Zeilinger, "On the Interpretation and Philosophical Foundation of Quantum Mechanics."

114 *not a feminist bank teller*: Kahneman, *Thinking, Fast and Slow*, 157–59.

114 *arrive at a contradiction*: Lapkiewicz et al., "Experimental Non-Classicality of an Indivisible Quantum System."

114 *"the main culprit"*: Zeilinger, "Testing Concepts of Reality with Entangled Photons in the Laboratory and Outside."

115 *"something different"*: Maudlin, "Special Relativity and Quantum Entanglement: How Compatible Are They?"

115 *"He derived it"*: Ibid.

115 *rederive his theorem*: Bell, *Speakable and Unspeakable in Quantum Mechanics*, 150, 157n10.

116 *they deny any need*: Zeilinger, "On the Interpretation and Philosophical Foundation of Quantum Mechanics."

116 *"you don't need nonlocality"*: Zeilinger, "Testing Concepts of Reality with Entangled Photons."

116 *"get to the bottom of it"*: Albert, "Physics and Narrative."

116 *"your money's worth"*: David Z. Albert, interview by author, March 16, 2011, Dresden, Germany.

116 *"very brilliant people"*: Jon P. Jarrett, e-mail to author, July 11, 2011.

117 *"Annual income twenty pounds"*: Dickens, *The Personal History of David Copperfield*, 231.

117 *"it was stuffy"*: Zeilinger, interview by author.

118 *combined with ordinary light or radio signals*: Brassard, "Quantum Communication Complexity (A Survey)."

118 *the nonlocal link*: Mattle et al., "Dense Coding in Experimental Quantum Communication."

118 *break the entanglement*: Jennewein et al., "Quantum Cryptography with Entangled Photons."

118 *"secure" or "blind" computation*: Barz et al., "Demonstration of Blind Quantum Computing."

118 *in a scrambled form*: He, "Simple Quantum Protocols for the Millionaire Problem with a Semi-Honest Third Party."

119 *"disbelieve all facts and theories"*: James, *The Will to Believe and Other Essays in Popular Philosophy*, 10.

119 *"not totally amazed"*: Nicolas Gisin, interview by author, November 8, 2010, Geneva.

119 *"steer" its partner*: Schrödinger, "Discussion of Probability Relations Between Separated Systems"; Wiseman, Jones, and Doherty, "Steering, Entanglement, Nonlocality, and the Einstein-Podolsky-Rosen Paradox."

119 *can remain muted*: Popescu, "Bell's Inequalities and Density Matrices: Revealing 'Hidden' Nonlocality."

119 *evading government surveillance*: Branciard et al., "One-Sided Device-Independent Quantum Key Distribution: Security, Feasibility, and the Connection with Steering."

120 *envisioned "superquantum" coins*: Rohrlich and Popescu, "Nonlocality as an Axiom for Quantum Theory."

120 *an available time slot*: Brassard et al., "A Limit on Nonlocality in Any World in Which Communication Complexity Is Not Trivial."

121 *an all-pervading fluid*: Bohm and Vigier, "Model of the Causal Interpretation of Quantum Theory in Terms of a Fluid with Irregular Fluctuations."

121 *high but still finite speed*: Bancal et al., "Quantum Non-Locality Based on Finite-Speed Causal Influences Leads to Superluminal Signalling."

121 *already at its destination*: Van Fraassen, *Quantum Mechanics: An Empiricist View*, 351.

121 *flashing warning light*: Bohm and Hiley, *The Undivided Universe: An Ontological Interpretation of Quantum Theory*, 42, 350, 374–78.

122 *"I'm asking myself"*: 't Hooft, interview by author, June 4, 2011, New York.

123 *effectively nonlocal*: Price, *Time's Arrow and Archimedes' Point*, 223–24.

123 *a breed of nonlocality*: Allori et al., "Many Worlds and Schrödinger's First Quantum Theory," 13–15; Wallace, *The Emergent Multiverse*, 303–10.

123 *to be an individual*: Howard, "A Peek Behind the Veil of Maya."

123 *share all your memories*: Tegmark, "Parallel Universes."

123 *Which one of them is you?*: Wallace, *The Emergent Multiverse*, 275–76.

123 *no explanation is possible*: Van Fraassen, *Quantum Mechanics: An Empiricist View*, 372–74.

123 *perhaps nothing causes*: Fine, "Do Correlations Need to Be Explained?"

123 *"belief in determinism"*: Fine, e-mail to author, April 27, 2011.

124 *"We have randomness"*: Gisin, interview by author.

124 *a failure of either aspect*: Einstein, "Autobiographical Notes," 85.

124 *go together in reality*: Jones and Clifton, "Against Experimental Metaphysics"; Cushing, "Locality/Separability: Is This Necessarily a Useful Distinction?" 111; Spekkens, "The Paradigm of Kinematics and Dynamics Must Yield to Causal Structure," 4.

125 *an empty statement*: Shimony, "Aspects of Nonlocality in Quantum Mechanics," 119; Ismael, "What Entanglement Might Be Telling Us," 14.

125 *loses none of its power*: Maudlin, *Quantum Non-Locality and Relativity*, 22–24; Lange, *An Introduction to the Philosophy of Physics: Locality, Fields, Energy, and Mass*, 281.

125 *"emerging in an organic way"*: Jenann Ismael, Skype interview by author, April 9, 2014.

126 *a kind of union*: Minkowski, "Raum Und Zeit."

126 *handled with care*: Healey, "Holism and Nonseparability."

126 *alternative-medicine practitioners*: Maudlin, "Part and Whole in Quantum Mechanics," 55.

126 *"from outside spacetime"*: Gisin, interview by author.

126 *"a fundamental failing"*: Maudlin, "Part and Whole," 55–56.

127 *"no story in spacetime"*: Gisin, interview by author.

5. NONLOCALITY AND THE UNIFICATION OF PHYSICS

129 *"streams of climbers"*: Giddings, e-mail to author, March 7, 2014.

130 *"community wasn't quite ready"*: Giddings, e-mail to author, July 4, 2013.

130 *implicitly assumed that forces leap*: Strocchi, "Relativistic Quantum Mechanics and Field Theory," 508.

130 *"this wonderful theory"*: Hertz, *Miscellaneous Papers*, 318.

131 *a big tangle*: Ashtekar and Rovelli, "A Loop Representation for the Quantum Maxwell Field."

131 *"formerly known as strings"*: Duff, "The Theory Formerly Known as Strings."

131 *"AdS/CFT duality"*: Horowitz and Polchinski, "Gauge/Gravity Duality."

131 *"the birth and growth"*: Donald Marolf, e-mail to author, June 17, 2013.

132 *"in the Alaska twilight"*: Giddings, e-mail to author, March 7, 2014.

132 *couldn't explain light*: Weinberg, "The Search for Unity: Notes for a History of Quantum Field Theory," 21.

132 *ultimately particle or wave*: Schweber, *QED and the Men Who Made It: Dyson, Feynman, Schwinger, and Tomonaga*, xxvi, 1–2, 37.

132 *created and annihilated*: Teller, *An Interpretive Introduction to Quantum Field Theory*, 82.

132 *big effervescent gobs*: Ibid., 112–13.

133 *not motes of matter*: Ibid., 69–81.

133 *case of convergent evolution*: Schweber, *QED and the Men Who Made It*, 53.

133 *As Pauli pointed out*: Blum, "From the Necessary to the Possible: The Genesis of the Spin-Statistics Theorem," 553–54.

133 *chops space into two parts*: Pauli, "The Connection Between Spin and Statistics," 721; Teller, *An Interpretive Introduction to Quantum Field Theory*, 83–84.

133 *alternatives also divvy up space*: Wright, "Quantum Field Theory: Motivating the Axiom of Microcausality."

134 *often deeply doubtful*: Weinberg, "The Search for Unity," 24–26; Schweber, *QED and the Men Who Made It*, 83–85.

134 *retook his first course*: Joseph Polchinski, e-mail to author, July 8, 2013.

134 *scared most of them off*: Teller, *Interpretive Introduction*, vii.

135 *"what the mathematics means"*: Hans P. Halvorson, interview by author, July 5, 2013, Princeton, NJ.

135 *two different places at once*: Teller, *Interpretive Introduction*, 86–87, 109; Halvorson and Clifton, "No Place for Particles in Relativistic Quantum Theories?"; Duncan, *The Conceptual Framework of Quantum Field Theory*, 160–63.

135 *banished nonlocality, not entrenched it*: Halvorson, "Locality, Localization, and the Particle Concept: Topics in the Foundations of Quantum Field Theory," 169–70.

136 *"gives rise to nonlocality"*: Halvorson, interview by author.

136 *only if relativity did not fully apply*: Newton and Wigner, "Localized States for Elementary Systems"; Teller, *Interpretive Introduction*, 56, 85–89.

136 *see it suddenly leap*: Ruijsenaars, "On Newton-Wigner Localization and Superluminal Propagation Speeds"; Hegerfeldt, "Violation of Causality in Relativistic Quantum Theory?"; Malament, "In Defense of Dogma: Why There Cannot Be a Relativistic Quantum Mechanics of (Localizable) Particles."

136 *differing answers*: Halvorson, "Locality, Localization, and the Particle Concept," 64; Kuhlmann, "What Is Real?"

136 *"There isn't anything"*: Halvorson, interview by author.

136 *on a guitar string*: Auyang, *How Is Quantum Field Theory Possible?*, 51–53, 157–60.

136 *waves are so jumbled*: Fraser, "The Fate of 'Particles' in Quantum Field Theories with Interactions."

137   *break down a big problem*: Teller, *Interpretive Introduction*, 139–42.

137   *"anything about reality"*: Halvorson, interview by author.

137   *conveyor belts and rolling drums*: Morus, *When Physics Became King*, 81–84.

138   *can't be an array of pixels*: Teller, *Interpretive Introduction*, 80–81, 98–99.

138   *also rule out pixels*: Halvorson, "Algebraic Quantum Field Theory," 778–79; Baker, "Against Field Interpretations of Quantum Field Theory."

138   *"seems like it's backfired"*: Halvorson, interview by author.

139   *don't convey information*: Peskin and Schroeder, *An Introduction to Quantum Field Theory*, 27–29.

139   *version of the spooky synchrony*: Wald, "Correlations and Causality in Quantum Field Theory," 300; Reznik, "Distillation of Vacuum Entanglement to EPR Pairs"; Franson, "Generation of Entanglement Outside of the Light Cone."

139   *points in the field are*: Summers and Werner, "The Vacuum Violates Bell's Inequalities"; Clifton et al., "Superentangled States."

139   *"constrained in momentum"*: Halvorson, interview by author.

139   *"superentanglement"*: Clifton et al., "Superentangled States."

139   *points lying beyond*: Wald, "Correlations and Causality in Quantum Field Theory," 301.

139   *sever the bonds*: Clifton and Halvorson, "Entanglement and Open Systems in Algebraic Quantum Field Theory."

139   *residual random jittering*: Summers and Werner, "The Vacuum Violates Bell's Inequalities"; Redhead, "More Ado About Nothing."

139   *"You may have forgotten"*: Halvorson, interview by author.

140   *despite various proposals*: Retzker, Cirac, and Reznik, "Detecting Vacuum Entanglement in a Linear Ion Trap"; Sabín et al., "Dynamics of Entanglement via Propagating Microwave Photons."

140   *a smidge farther apart*: Summers and Werner, "The Vacuum Violates Bell's Inequalities," 258–59; Halvorson, "Locality, Localization, and the Particle Concept," 18.

140   *"nonlocality at all distances"*: Halvorson, interview by author.

140   *fields may become*: Reznik, "Distillation of Vacuum Entanglement to EPR Pairs."

140   *entangled with each other*: Haroche and Raimond, *Exploring the Quantum: Atoms, Cavities, and Photons*, 285.

140   *a Bollywood movie*: Wen, "Topological Order: From Long-Range Entangled Quantum Matter to a Unified Origin of Light and Electrons."

140   *that once seemed magical*: Sachdev, "Strange and Stringy."

140   *"Laypeople love it"*: Arkani-Hamed, interview by author, September 24, 2013, Princeton, NJ.

141   *"think of mutual entanglement"*: Arkani-Hamed, interview by author, March 29, 2012, Pasadena, CA.

142   *Faraday conducted electrical experiments*: Faraday, *Experimental Researches in Electricity*, vol. 1, 83–84, 364–66; Healey, *Gauging What's Real: The Conceptual Foundations of Contemporary Gauge Theories*, 4, 155–57.

142   *"lighted candles, electrometers"*: Faraday, *Experimental Researches in Electricity*, vol. 1, 366.

142   *mathematical complications*: Hatfield, *Quantum Field Theory of Point Particles and Strings*, 77–81.

142 *"a complete fiction"*: Arkani-Hamed, "Space-Time, Quantum Mechanics and Scattering Amplitudes."

142 *"the first of many, many hints"*: Arkani-Hamed, interview by author, September 24, 2013.

143 *fudge factor*: Belot, "Understanding Electromagnetism," 541–42; Healey, *Gauging What's Real*, 25–26, 49.

143 *fix the value*: Bork, "Maxwell and the Vector Potential."

143 *disturbances in the potential*: Jackson, *Classical Electrodynamics*, 222–23.

143 *to "murder" it*: Hunt, *The Maxwellians*, 115–18, 165–66.

144 *in the argot*: Castellani, "Dirac on Gauges and Constraints."

144 *Because of the constraint*: Jackson, *Classical Electrodynamics*, 271.

144 *the constraint is defined*: Strocchi, "Gauss' Law in Local Quantum Field Theory," 229–31; Ashtekar and Rovelli, "A Loop Representation for the Quantum Maxwell Field," 1149.

144 *freedom from such constraints*: Earman, "Locality, Nonlocality, and Action at a Distance: A Skeptical Review of Some Philosophical Dogmas," 457, 458.

145 *Electrons, rather than lighted candles*: Aharonov and Bohm, "Significance of Electromagnetic Potentials in the Quantum Theory."

145 *tricky to do*: Matteucci and Pozzi, "New Diffraction Experiment on the Electrostatic Aharonov-Bohm Effect."

145 *analogous test for magnetism*: Ibid.; Chambers, "Shift of an Electron Interference Pattern by Enclosed Magnetic Flux."

145 *this came as a shock*: Ramsey, *Spectroscopy with Coherent Radiation: Selected Papers of Norman F. Ramsey*, 399.

145 *localized structures are mismatched*: Wu and Yang, "Concept of Nonintegrable Phase Factors and Global Formulation of Gauge Fields."

145 *"high school physics"*: Richard A. Healey, interview by author, March 22, 2014, Irvine, CA.

146 *"I felt very lonely"*: Healey, "New Thoughts on Yang-Mills Theories."

146 *more as spectators*: Healey, *Gauging What's Real*, 31, 53, 127; Wallace, *The Emergent Multiverse*, 294n7.

146 *"People draw distinctions"*: Healey, interview by author.

146 *separable almost by definition*: Howard, "Holism, Separability and the Metaphysical Implications of the Bell Experiments," 232–40.

146 *goes back to Dirac*: Dirac, "Quantised Singularities in the Electromagnetic Field"; Mandelstam, "Quantum Electrodynamics Without Potentials"; Wu and Yang, "Concept of Nonintegrable Phase Factors."

147 *"Loops, not points"*: Healey, e-mail to author, July 24, 2013.

147 *satisfying local action*: Lyre, "Holism and Structuralism in $U(1)$ Gauge Theory," 657–60.

147 *left with some ambiguity*: Rozali, "Comments on Background Independence and Gauge Redundancies."

147 *"not losing the points"*: Healey, interview by author.

148 *"a question about here"*: Marolf, interview by author, October 12, 2010, Santa Barbara, CA.

148 *resurvey tectonic zones*: Musser, "What Happens to Google Maps When Tectonic Plates Move?"

149 *Earth's mass warps time*: Unruh, "Time, Gravity, and Quantum Mechanics"; Schutz, *Gravity from the Ground Up: An Introductory Guide to Gravity and General Relativity*, 229–32.

149 *the warping of space*: Schutz, *Gravity from the Ground Up*, 234–36.

150 *a mini-wormhole*: Howard, "Holism, Separability, and the Metaphysical Implications of the Bell Experiments," 251–52; Maldacena and Susskind, "Cool Horizons for Entangled Black Holes."

150 *random and unaccountable things*: Coleman, "Why There Is Nothing Rather Than Something: A Theory of the Cosmological Constant"; Markopoulou and Smolin, "Quantum Theory from Quantum Gravity."

150 *overlap with itself*: Misner and Wheeler, "Classical Physics as Geometry," 552.

150 *To foil such paradoxes*: Horwich, *Asymmetries in Time: Problems in the Philosophy of Science*, 124–25; Friedman et al., "Cauchy Problem in Spacetimes with Closed Timelike Curves."

151 *constraint is a form of nonlocality*: Earman, "Recent Work on Time Travel"; Arntzenius and Maudlin, "Time Travel and Modern Physics."

151 *no external or absolute standard*: Misner, "Feynman Quantization of General Relativity," 499; Misner, Thorne, and Wheeler, *Gravitation*, 429–31; Rovelli, *Quantum Gravity*, 65–75.

151 *"honestly, Einstein didn't understand"*: Marolf, interview by author, October 12, 2010.

151 *shape of space would become ambiguous*: Stachel, *Einstein from 'B' to 'Z,'* 323–25.

152 *"in the reshuffled spacetime"*: Marolf, e-mail to author, June 26, 2013.

152 *"to the whole spacetime"*: Marolf, interview by author, March 26, 2012, Santa Barbara, CA.

152 *version of gauge invariance*: Kuchař, "Time and Interpretations of Quantum Gravity," 2–3.

153 *lack any differentiating attributes*: Stachel, *Einstein from 'B' to 'Z,'* 312–18.

153 *Those quantities must be holistic*: Misner, Thorne, and Wheeler, *Gravitation*, 466–68; Rickles, *Symmetry, Structure, and Spacetime*, 129–31.

153 *"Any theory of gravity"*: Marolf, "Discussion: Holography and Unitarity in Black Hole Evaporation."

153 *some independent reality*: Earman, *World Enough and Space-Time: Absolute Versus Rational Theories of Space and Time*, 96–108, Greene, *The Fabric of the Cosmos*, 73–75.

153 *anchor a coordinate grid*: Rovelli, *Quantum Gravity*, 88–96.

154 *"in a continuous way"*: Marolf, interview by author, March 26, 2012.

154 *still only "pseudo-local"*: Giddings, Marolf, and Hartle, "Observables in Effective Gravity."

154 *how strong it will be*: Penrose, *The Road to Reality: A Complete Guide to the Laws of the Universe*, 841.

155 *"we would never know"*: Gibbons, "Black Holes and Information."

155 *holistic feature of spacetime*: Lam, "Structural Aspects of Space-Time Singularities."

155 *having an edge*: Harrison, *Cosmology: The Science of the Universe*, 149–53.

155 *toyed with hypothetical models*: Randall and Sundrum, "A Large Mass Hierarchy from a Small Extra Dimension."

156 *fixes the shape of space*: Marolf, "Holographic Thought Experiments," 2.

156 *"nailed to the boundary"*: Marolf, e-mail to author, February 24, 2014.

156 *Locality on the boundary*: Marolf, "Unitarity and Holography in Gravitational Physics"; Marolf, "Holography Without Strings?"

156 *"equal to bulk observables"*: Marolf, e-mail to author, October 17, 2013.

157 *can't be as fundamental*: Greene, *The Fabric of the Cosmos*, 477.

158 *the dissolution of space*: Gorelik, "Matvei Bronstein and Quantum Gravity: 70th Anniversary of the Unsolved Problem."

158 *limit of the reductionist program*: Seiberg, "Emergent Spacetime," 176.

159 *Einstein wrote to a friend*: Stachel, *Einstein from 'B' to 'Z,'* 149.

159 *spontaneously on all scales*: Wilczek, "Quantum Field Theory," S87–S88.

159 *the continuum must crumble*: Beller, *Quantum Dialogue: The Making of a Revolution*, 19–22.

159 *"lattice world"*: Carazza and Kragh, "Heisenberg's Lattice World: The 1930 Theory Sketch."

159 *come into play on small scales*: Zee, *Quantum Field Theory in a Nutshell*, chap. 8.3.

159 *no amount of rounding out*: Rovelli, *Quantum Gravity*, 404.

160 *threatens to disintegrate*: Markopoulou, "Space Does Not Exist, So Time Can," 4.

160 *like a balloon*: Witten, "Reflections on the Fate of Spacetime"; Martinec, "Evolving Notions of Geometry in String Theory."

160 *"simply no longer valid"*: Polchinski, interview by author, October 13, 2010, Santa Barbara, CA.

160 *also in the macroverse*: Markopoulou and Smolin, "Disordered Locality in Loop Quantum Gravity States."

161 *"something very small"*: Markopoulou, telephone interview by author.

161 *miniature tuning forks*: O'Connell et al., "Quantum Ground State and Single-Phonon Control of a Mechanical Resonator."

162 *"long-distance issues"*: Giddings, interview by author, May 14, 2012.

162 *an uneasy silence*: Redhead, "More Ado About Nothing."

162 *does not, in fact, think they cancel out*: Teller, *An Interpretive Introduction to Quantum Field Theory*, 110–11.

162 *draining the hole*: Jacobson, "Introduction to Quantum Fields in Curved Spacetime and the Hawking Effect," 23–27.

163 *not a vacuum*: Mathur, "The Information Paradox: A Pedagogical Introduction"; Almheiri et al., "Black Holes: Complementarity or Firewalls?"

163 *balances both needs*: Hossenfelder, "Disentangling the Black Hole Vacuum," 258.

163 *"a violent breakdown"*: Polchinski, interview by author.

163 *doubling the radius*: 't Hooft, "Dimensional Reduction in Quantum Gravity," 288.

164 *entropy increases fourfold*: Bekenstein, "Black Holes and Entropy."

164 *squeezing hard enough*: Ibid., 2339.

164 *boundary and volume are equivalent*: Rickles, "AdS/CFT Duality and the Emergence of Spacetime."

164 *the boundary is the fundamental reality*: Bousso, "The Holographic Principle," 859–60.

165 *"it's still very obscure"*: Daniel Kabat, interview by author, April 13, 2012. Bits, Branes, Black Holes, Santa Barbara, CA.

166 *In Maldacena's approach*: Seiberg, "Emergent Spacetime," 171–72; Horowitz and Polchinski, "Gauge/Gravity Duality."

166 *gave Maldacena a lift*: Alex Maloney, interview by author, March 23, 2012. Bits, Branes, Black Holes, Santa Barbara, CA.

6. SPACETIME IS DOOMED

167 *"didn't tell my sister"*: Ismael, e-mail to author, May 20, 2015.
167 *"a piece of glass"*: Ismael, Skype interview by author.
168 *metaphor for nonlocality*: Ismael, "What Entanglement Might Be Telling Us," 1–2.
168 *"different parts of space"*: Ibid.
168 *"the fundamental level"*: Michael Heller, interview by author, March 13, 2008, New York.
169 *prime matter*: Guthrie, *A History of Greek Philosophy*, vol. 2, *The Presocratic Tradition from Parmenides to Democritus*, 143.
169 *rearrangements of atoms*: Leucippus and Democritus, *The Atomists, Leucippus and Democritus*, 85n78; Lucretius, *The Nature of Things*, 7–11, 25–27.
169 *"something more basic"*: Arkani-Hamed, telephone interview by author.
169 *"talk about emergent spacetime"*: Arkani-Hamed, telephone interview by author.
169 *the realm beyond space*: Stump and Kretzmann, "Eternity"; Heller, "Where Physics Meets Metaphysics," 263–64.
170 *step toward emergent spacetime*: Slowik, "The Deep Metaphysics of Quantum Gravity: The Seventeenth Century Legacy and an Alternative Ontology Beyond Substantivalism and Relationism."
170 *"to breathe in empty space"*: Einstein, "Physics and Reality," 378.
170 *built out of "pregeometry"*: Misner, Thorne, and Wheeler, *Gravitation*, 1212.
170 *"outside our usual language"*: Arkani-Hamed, telephone interview by author.
170 *"For 2,000-plus years"*: Arkani-Hamed, "Space-Time, Quantum Mechanics, and the Large Hadron Collider."
170 *what it'll do in the future*: Earman, *A Primer on Determinism*, 4–7, 30–31.
170 *Rube Goldberg contraptions*: OK Go, "This Too Shall Pass."
171 *glue that binds the universe*: Barbour, *The End of Time: The Next Revolution in Physics*, 18.
171 *limited by the spatial arrangement*: Lieb and Robinson, "The Finite Group Velocity of Quantum Spin Systems."
172 *"interactive distance"*: Albert, "Elementary Quantum Metaphysics."
172 *"the lion is close"*: Albert, interview by author, April 18, 2011, New York.
173 *one mystery for another*: Maudlin, "Buckets of Water and Waves of Space: Why Spacetime Is Probably a Substance," 192–94, 196.
173 *the "inverse problem"*: Smolin, *Time Reborn: From the Crisis in Physics to the Future of the Universe*, 184–86.
173 *"Where spacetime exists"*: Moshe Rozali, telephone interview by author, July 13, 2011.
173 *time isn't emergent*: Smolin, *Time Reborn*; Maudlin, *New Foundations for Physical Geometry*.
173 *a powerful organizing role*: Musser, "Could Time End?"
173 *turned this thinking upside down*: Leibniz, *Philosophical Papers and Letters: A*

Selection, vol. 2, 666–74; Mehlberg, *Time, Causality, and the Quantum Theory*, vol. 1, 42–50.

174 *map out the entire world*: Bombelli et al., "Space-Time as a Causal Set."

174 *the number of such atoms*: Riemann, "On the Hypotheses Which Lie at the Bases of Geometry," 37.

174 *relations must be highly ordered*: Dowker, "Causal Sets and the Deep Structure of Spacetime," 454; Henson, "The Causal Set Approach to Quantum Gravity," 405.

174 *"nothing like spacetime"*: Dowker, interview by author.

175 *they are all interlocked*: Barbour, "The Nature of Time."

175 *"the interdependence of things"*: Mach, *The Science of Mechanics: A Critical and Historical Account of Its Development*, 223, 224.

175 *the odometer mileage*: Michael Mouser, telephone interview by author, June 3, 2014.

176 *The distances obey*: Lefschetz, *Introduction to Topology*, 28.

177 *lose several more numbers*: Einstein, *The Meaning of Relativity*, 8; Weinberg, *Gravitation and Cosmology: Principles and Applications of the General Theory of Relativity*, 6–8; Misner, Thorne, and Wheeler, *Gravitation*, 306–308.

177 *"It's data compression"*: Julian B. Barbour, interview by author, May 22, 2010, at Laws of Nature: Their Nature and Knowability, conference, Perimeter Institute for Theoretical Physics, Waterloo, Ontario, Canada, May 20–22, 2010.

179 *"unmediated" distances*: Nerlich, *The Shape of Space*, 18–23.

181 *struggling to follow* Game of Thrones: McGoldrick, Gerson, and Petry, *Genograms: Assessment and Intervention*.

181 *This kind of self-organizing*: Smith and Foley, "Classical Thermodynamics and Economic General Equilibrium Theory."

182 *relationships become regimented*: Markopoulou, "Space Does Not Exist, So Time Can," 7–8.

182 *mind-bogglingly complex*: Rovelli, "'Forget Time.'"

182 *space as a network*: Misner, Thorne, and Wheeler, *Gravitation*, 1203–12; Bohm and Hiley, *The Undivided Universe: An Ontological Interpretation of Quantum Theory*, 374–78; Penrose, *The Road to Reality: A Complete Guide to the Laws of the Universe*, 946–50; Finkelstein, "Space-Time Code."

182 *trying to make the idea work*: Kaplunovsky and Weinstein, "Space-Time: Arena or Illusion?"; Meschini, Lehto, and Piilonen, "Geometry, Pregeometry and Beyond."

182 *inject a sense of humor*: Konopka, Markopoulou, and Severini, "Quantum Graphity: A Model of Emergent Locality."

183 *from earthquakes to ecosystems*: Stanley et al., "Scale Invariance and Universality: Organizing Principles in Complex Systems."

183 *"very ambitious and dangerous"*: Claus Kiefer, interview by author, August 31, 2011, Copenhagen.

184 *"just put out your hand"*: Markopoulou and Kuhn, "Why Is the Universe So Breathtaking?"

185 *passage of the signal*: Hamma and Markopoulou, "Background-Independent Condensed Matter Models for Quantum Gravity."

185 *"cool the universe"*: Markopoulou, interview by author.

185 *the primordial grains*: Hamma and Markopoulou, "Background-Independent Condensed Matter Models," 13–16.

185 *"causal dynamical triangulations"*: Loll, Ambjørn, and Jurkiewicz, "The Universe from Scratch."

185 *jelly-bean experiment*: Surowiecki, *The Wisdom of Crowds: Why the Many Are Smarter Than the Few and How Collective Wisdom Shapes Business, Economies, Societies, and Nations*, 5.

186 *errors offset one another*: Ibid., 10–11, 27–30.

186 *but the twisty paths*: Feynman, *QED: The Strange Theory of Light and Matter*, 53–54.

186 *best-known matrix model*: Banks et al., "M Theory as a Matrix Model: A Conjecture."

187 *"bunch of Tinkertoy parts"*: Leonard Susskind, interview by author, February 18, 2011, Palo Alto, CA.

187 *the endpoints of strings*: Musser, *The Complete Idiot's Guide to String Theory*, 155.

188 *its constituents' self-interactions*: Banks, "The State of Matrix Theory," 342–43; Martinec, "Evolving Notions of Geometry in String Theory," 167–68.

188 *"just a cancellation"*: Susskind, interview by author.

189 *"not the behavior we expect"*: Emil J. Martinec, e-mail to author, May 10, 2014.

191 *the scale of waves*: Susskind and Witten, "The Holographic Bound in Anti–de Sitter Space"; Balasubramanian and Kraus, "Spacetime and the Holographic Renormalization Group."

191 *a dramatic way of putting it*: Sundrum, "From Fixed Points to the Fifth Dimension."

192 *"drawing the Washington Monument"*: Raman Sundrum, Skype interview by author, October 3, 2014.

192 *A Chinese gong*: Fletcher, "Nonlinear Dynamics and Chaos in Musical Instruments."

192 *propagating through scale*: Balasubramanian et al., "Holographic Probes of Anti–de Sitter Spacetimes"; Heemskerk et al., "Holography from Conformal Field Theory."

192 *stitching space together*: Mark Van Raamsdonk, interview by author, September 30, 2010, Vancouver, B.C.

192 *creates space*: Nishioka, Ryu, and Takayanagi, "Holographic Entanglement Entropy: An Overview"; Van Raamsdonk, "Building Up Spacetime with Quantum Entanglement"; Swingle, "Constructing Holographic Spacetimes Using Entanglement Renormalization."

193 *"disordered locality"*: Markopoulou and Smolin, "Disordered Locality in Loop Quantum Gravity States."

193 *the term "long links"*: Brian Swingle, e-mail to author, August 22, 2014.

194 *to explain particles*: Einstein and Rosen, "The Particle Problem in the General Theory of Relativity."

194 *join entangled particles*: Jensen and Karch, "Holographic Dual of an Einstein-Podolsky-Rosen Pair Has a Wormhole"; Baez and Vicary, "Wormholes and Entanglement."

194 *"closeness through the wormhole"*: Juan Maldacena, interview by author, September 17, 2013, Princeton, NJ.

194 *like overkill*: Maudlin, *Quantum Non-Locality and Relativity: Metaphysical Intimations of Modern Physics*, 238.

194   *pinches off the wormhole*: Maldacena and Susskind, "Cool Horizons for Entangled Black Holes," 782.

194   *so inordinately difficult*: Swingle, Skype interview by author, August 21, 2014.

194   *a gibberish one*: Markopoulou and Smolin, "Disordered Locality," 3822; Valentini, "Beyond the Quantum," 36.

195   *"engulfed by globality"*: Heller, interview by author.

195   *philosophers and sociologists*: Pettit, *The Common Mind: An Essay on Psychology, Society, and Politics*, 166–73.

195   *sensitive to what happens*: Heller and Sasin, "Einstein-Podolski-Rosen Experiment from Noncommutative Quantum Gravity"; Heller and Sasin, "Nonlocal Phenomena from Noncommutative Pre-Planckian Regime."

195   *set the cosmos in motion*: Misner, Thorne, and Wheeler, *Gravitation*, 1196–97.

196   *shaped itself*: Craps, Sethi, and Verlinde, "A Matrix Big Bang"; Martinec, Robbins, and Sethi, "Toward the End of Time."

196   *ancient creation myths*: Casey, *The Fate of Place: A Philosophical History*, 9–16.

196   *"ultralocality"*: Carlip, "Spontaneous Dimensional Reduction?"; Mielczarek, "Asymptotic Silence in Loop Quantum Cosmology."

196   *would have sloshed around*: Hamma and Markopoulou, "Background-Independent Condensed Matter Models," 12; McFadden and Skenderis, "Observational Signatures of Holographic Models of Inflation."

196   *"a completely connected graph"*: Markopoulou, e-mail to author.

197   *A phase transition*: Easther et al., "Constraining Holographic Inflation with WMAP"; Dreyer, "The World Is Discrete."

197   *a fluid turmoil*: Witten, "Anti–de Sitter Space, Thermal Phase Transition, and Confinement in Gauge Theories."

197   *"conventional geometric notion"*: Martinec, e-mail to author, April 23, 2010.

197   *effectively infinite-dimensional*: Sekino and Susskind, "Fast Scramblers."

197   *finding the exit*: Hamma and Markopoulou, "Background-Independent Condensed Matter Models for Quantum Gravity," 16–17.

197   *"acts as a trap"*: Markopoulou, interview by author.

198   *the black hole disperses*: Banks et al., "Schwarzschild Black Holes from Matrix Theory," 229; Horowitz, Lawrence, and Silverstein, "Insightful D-Branes," 18–20.

198   *"this huge space of states"*: Martinec, e-mail to author, August 12, 2014.

199   *"eating up spacetime"*: Rozali, telephone interview by author.

199   *becomes too disordered*: Balasubramanian, "What We Don't Know About Time," 109–111.

199   *in four dimensions*: Sachdev, "Strange and Stringy."

199   *an inner simplicity*: Nastase, "The RHIC Fireball as a Dual Black Hole"; Horowitz and Polchinski, "Gauge/Gravity Duality," 181.

200   *preprogrammed into the rules*: Nerlich, *The Shape of Space*, chap. 1; Meschini, Lehto, and Piilonen, "Geometry, Pregeometry and Beyond."

200   *"What are the rules?"* Martinec, interview by author, June 18, 2014, Chicago.

200   *"Space is totally overrated"*: Carroll, "Setting Time Aright."

200   *emergence without presupposing time*: Heller and Sasin, "Emergence of Time"; Balasubramanian, "What We Don't Know About Time"; Aoki et al., "Space-Time Structures from IIB Matrix Model."

200   *the emergent dimension is temporal*: Strominger, "Inflation and the dS/CFT Correspondence."

CONCLUSION: THE AMPLITUHEDRON

203 *controversial visit to Copenhagen*: Powers, *Heisenberg's War: The Secret History of the German Bomb*, 120–28.

203 *American spy*: Ibid., 294–97.

203 *Berg let him live*: Ibid., 394–401.

204 *S-matrix tabulates*: Cushing, *Theory Construction and Selection in Modern Physics: The S Matrix*, 32–34; Schweber, *QED and the Men Who Made It: Dyson, Feynman, Schwinger, and Tomonaga*, 154–55.

204 *do full-up calculations*: Weinberg, "The Search for Unity: Notes for a History of Quantum Field Theory," 26–30.

204 *the S-matrix regained its appeal*: Cushing, *Theory Construction and Selection*, 115–18.

204 *all there is*: Ibid., 142–45.

205 *"such an unobservable continuum"*: Chew, "The Dubious Role of the Space-Time Continuum in Subatomic Physics," 529.

205 *nothing is more fundamental*: Capra, *The Tao of Physics: An Exploration of the Parallels Between Modern Physics and Eastern Mysticism*, 286–87.

205 *like sudoku*: Chew, "'Bootstrap': A Scientific Idea?" 763–64.

205 *on macroscopic scales*: Stapp, "Space and Time in S-Matrix Theory"; Capra, *The Tao of Physics*, 318.

205 *"nature consistent with itself"*: Ibid., 762.

205 *no unique S-matrix*: Gross, "Twenty Five Years of Asymptotic Freedom," 429.

205 *expanded the framework*: Penrose, *The Road to Reality: A Complete Guide to the Laws of the Universe*, 962–66.

205 *more fundamental than local ones*: Ibid., 963, 991.

206 *the word "twistor"*: Ibid., 980–82.

206 *"are encoded nonlocally"*: Lionel Mason, interview by author, March 14, 2010, Oxford.

206 *"Line-of-sight ideas"*: Andrew Hodges, interview by author, March 16, 2010, Oxford.

206 *to all those stars*: Penrose, *The Road to Reality*, 1049.

206 *"closer than yesterday"*: Rafael D. Sorkin, e-mail to author, November 16, 2007.

206 *"googly" problem*: Penrose, *The Road to Reality*, 1000.

206 *"He's very dogged"*: Mason, interview by author, March 14, 2010.

206 *"the ugly duckling"*: Arkani-Hamed, "Space-Time, Quantum Mechanics and Scattering Amplitudes."

206 *"excitement and depression"*: Halvorson, interview by author.

207 *"I'm discouraged more"*: Markopoulou, interview by author.

207 *"It's the rest of my life"*: Markopoulou, telephone interview by author.

207 *seventy-page monster*: Witten, "Perturbative Gauge Theory as a String Theory in Twistor Space."

207 *"fascinating and exciting"*: Roger Penrose, interview by author, March 15, 2010, Oxford.

207 *twistors in their heyday*: Witten, "An Interpretation of Classical Yang-Mills Theory."

207 *"basically thinking off and on"*: Edward Witten, e-mail to author, January 6, 2015.

208 *"this beautiful idea"*: Mason, interview by author, March 14, 2010.

208 *inputs and outputs*: Britto et al., "Direct Proof of the Tree-Level Scattering Amplitude Recursion Relation in Yang-Mills Theory."

208 *"'Revenge of the Analytic S-Matrix'"*: Lance J. Dixon, e-mail to author, February 16, 2010.

208 *"started vaguely looking like Andrew's pictures"*: Arkani-Hamed, telephone interview by author.

208 *geometrically using twistors*: Arkani-Hamed et al., "A Note on Polytopes for Scattering Amplitudes."

209 *a suitably funky name*: Arkani-Hamed and Trnka, "The Amplituhedron."

209 *"There are no fields"*: Jaroslav Trnka, e-mail to author, September 16, 2014.

209 *"Simple geometric properties"*: Arkani-Hamed, interview by author, September 24, 2013.

209 *"These building blocks"*: Ibid.

211 *Science is dispute*: Beller, *Quantum Dialogue*, 310; Freire, *The Quantum Dissidents*, 2–4.

211 *"all the blind alleys"*: Arkani-Hamed, interview by author, September 24, 2013.

211 *when challenged, doubling down*: Mitroff, "Norms and Counter-Norms in a Select Group of the Apollo Moon Scientists: A Case Study of the Ambivalence of Scientists," 588–89.

211 *a comfort level*: Traweek, *Beamtimes and Lifetimes: The World of High Energy Physicists*, 113–23.

211 *origins in wrong ideas*: Johnson, *Where Good Ideas Come From: The Natural History of Innovation*, 134–39.

212 *"Darwinian survival value"*: Susskind, *The Cosmic Landscape: String Theory and the Illusion of Intelligent Design*, 271.

213 *"they are children of their age"*: Schrödinger, *Science and the Human Temperament*, 80.

213 *"death of distance"*: Cairncross, *The Death of Distance: How the Communications Revolution Will Change Our Lives*.

213 *the boundaries of the self*: Thurschwell, *Literature, Technology and Magical Thinking, 1880–1920*, 12–14.

213 *during Hurricane Katrina*: Baum, "Deluged."

213 *within a few tens of miles*: Wijsman and Cavalli-Sforza, "Migration and Genetic Population Structure with Special Reference to Humans."

213 *college-educated people*: Taylor et al., "American Mobility: Who Moves? Who Stays Put? Where's Home?"

214 *meld into a mob*: Zimbardo, *The Lucifer Effect*, chap. 13.

214 *"the great uniters and dividers"*: Schlick, *General Theory of Knowledge*, 53.

214 *dark matter and dark energy*: Seiberg, "Emergent Spacetime," 167; Henson, "The Causal Set Approach to Quantum Gravity," 13; Prescod-Weinstein and Smolin, "Disordered Locality as an Explanation for the Dark Energy"; Verlinde, "The Dark Phase Space of de Sitter."

214 *under which we could travel faster than light*: Hashimoto and Itzhaki, "Traveling Faster Than the Speed of Light in Noncommutative Geometry"; Valentini, "Beyond the Quantum."

A NOTE ON ENTANGLEMENT

215 *need not be a "beam"*: Jaeger and Sergienko, "Multi-photon quantum interferometry."

217 *sandwich two thin layers*: Kwiat et al., "Ultrabright source of polarization-entangled photons."

217 *Before physicists discovered*: d'Espagnat, "The Quantum Theory and Reality."

217 *fluorescent light bulbs*: Edward Fry, e-mail to author, February 4, 2016.

Bibliography

Abiko, Seiya. "Einstein's Kyoto Address: 'How I Created the Theory of Relativity.'" *Historical Studies in the Physical and Biological Sciences* 31, no. 1 (2000): 1–35.

Ade, P.A.R., R. W. Aikin, D. Barkats, S. J. Benton, C. A. Bischoff, J. J. Bock, J. A. Brevik, et al. "Detection of B-Mode Polarization at Degree Angular Scales by BICEP2." *Physical Review Letters* 112 (June 20, 2014).

Agrippa von Nettesheim, Heinrich Cornelius. *Three Books of Occult Philosophy.* Edited by Donald Tyson. Translated by James Freake. St. Paul, MN: Llewellyn Worldwide, 1993.

Aharonov, Yakir, and David Bohm. "Significance of Electromagnetic Potentials in the Quantum Theory." *Physical Review* 115, no. 3 (August 1, 1959): 485–91.

Aharonov, Yakir, and Daniel Rohrlich. *Quantum Paradoxes.* Berlin: Wiley-VCH, 2005.

Albert, David Z. "Elementary Quantum Metaphysics." In *Bohmian Mechanics and Quantum Theory: An Appraisal*, edited by James T. Cushing, Arthur Fine, and Sheldon Goldstein, 277–84. Boston: Kluwer Academic Publishers, 1996.

———. "Physics and Narrative." Intersectional Symposium: The Concept of Reality in Physics, Dresden, Germany, March 16, 2011.

Allori, Valia, Sheldon Goldstein, Roderich Tumulka, and Nino Zanghì. "Many Worlds and Schrödinger's First Quantum Theory." *The British Journal for the Philosophy of Science* 62, no. 1 (March 2011): 1–27.

Almheiri, Ahmed, Donald Marolf, Joseph Polchinski, and James Sully. "Black Holes: Complementarity or Firewalls?" *Journal of High Energy Physics*, no. 2 (February 11, 2013): 1–20.

Aoki, Hajime, Satoshi Iso, Hikaru Kawai, Yoshihisa Kitazawa, and Tsukasa Tada. "Space-Time Structures from IIB Matrix Model." *Progress of Theoretical Physics* 99, no. 5 (May 1998): 713–45.

Aristotle. *Physics.* Edited by David Bostock. Translated by Robin Waterfield. New York: Oxford University Press, 1996.

Arkani-Hamed, Nima. "Space-Time, Quantum Mechanics and Scattering Amplitudes." Columbia University Department of Physics Colloquium, New York, March 28, 2011.

————. "Space-Time, Quantum Mechanics, and the Large Hadron Collider." Institute for Advanced Study Public Lecture, Princeton, NJ, February 23, 2011.

Arkani-Hamed, Nima, Jacob Bourjaily, Freddy Cachazo, and Jaroslav Trnka. "Local Spacetime Physics from the Grassmannian." *Journal of High Energy Physics*, no. 1 (January 24, 2011).

Arkani-Hamed, Nima, Jacob Bourjaily, Freddy Cachazo, Andrew Hodges, and Jaroslav Trnka. "A Note on Polytopes for Scattering Amplitudes." *Journal of High Energy Physics*, no. 4 (April 16, 2012).

Arkani-Hamed, Nima, Freddy Cachazo, and Jared Kaplan. "What Is the Simplest Quantum Field Theory?" *Journal of High Energy Physics*, no. 9 (September 6, 2010).

Arkani-Hamed, Nima, and Jaroslav Trnka. "The Amplituhedron." *Journal of High Energy Physics*, no. 10 (October 6, 2014).

Arntzenius, Frank, and Tim Maudlin. "Time Travel and Modern Physics." In *Time, Reality and Experience*, edited by Craig Callender, 169–200. Royal Institute of Philosophy Supplements. Cambridge: Cambridge University Press, 2002.

Ashtekar, Abhay, and Carlo Rovelli. "A Loop Representation for the Quantum Maxwell Field." *Classical and Quantum Gravity* 9, no. 5 (May 1992): 1121–50.

Auyang, Sunny Y. *How Is Quantum Field Theory Possible?* New York: Oxford University Press, 1995.

Bacciagaluppi, Guido. "Remarks on Space-Time and Locality in Everett's Interpretation." In *Non-Locality and Modality*, edited by Tomasz Placek and Jeremy Butterfield, 105–22. New York: Springer, 2002.

Bacciagaluppi, Guido, and Antony Valentini. *Quantum Theory at the Crossroads: Reconsidering the 1927 Solvay Conference.* New York: Cambridge University Press, 2009.

Baez, John C., and Jamie Vicary. "Wormholes and Entanglement." *arXiv.org*, January 14, 2014.

Baker, David John. "Against Field Interpretations of Quantum Field Theory." *The British Journal for the Philosophy of Science* 60, no. 3 (August 11, 2009): 585–609.

Balasubramanian, Vijay. "What We Don't Know About Time." *Foundations of Physics* 43, no. 1 (January 2013): 101–14.

Balasubramanian, Vijay, and Per Kraus. "Spacetime and the Holographic Renormalization Group." *Physical Review Letters* 83, no. 18 (November 1, 1999).

Balasubramanian, Vijay, Per Kraus, Albion Lawrence, and Sandip P. Trivedi. "Holographic Probes of Anti–de Sitter Spacetimes." *Physical Review D* 59, no. 10 (April 26, 1999).

Bancal, Jean-Daniel, Stefano Pironio, Antonio Acin, Yeong-Cherng Liang, Valerio Scarani, and Nicolas Gisin. "Quantum Non-Locality Based on Finite-Speed Causal Influences Leads to Superluminal Signalling." *Nature Physics* 8, no. 12 (December 2012): 867–70.

Banks, Tom. "The State of Matrix Theory." *Nuclear Physics B—Proceedings Supplements* 62, nos. 1–3 (March 1998): 341–47.

Banks, Tom, Willy Fischler, Igor R. Klebanov, and Leonard Susskind. "Schwarzschild Black Holes from Matrix Theory." *Physical Review Letters* 80, no. 2 (January 12, 1998): 226–29.

Banks, Tom, Willy Fischler, Stephen H. Shenker, and Leonard Susskind. "M Theory as a Matrix Model: A Conjecture." *Physical Review D* 55, no. 8 (April 15, 1997): 5112–28.

Banks, Tom, Leonard Susskind, and Michael E. Peskin. "Difficulties for the Evolution of Pure States into Mixed States." *Nuclear Physics B* 244, no. 1 (September 24, 1984): 125–34.

Barbour, Julian B. *Absolute or Relative Motion?* Vol. 1, *The Discovery of Dynamics*. New York: Cambridge University Press, 1989.

———. *The End of Time: The Next Revolution in Physics*. New York: Oxford University Press, 1999. Reprint, 2001.

———. "The Nature of Time." Essay written for the Foundational Questions Institute Essay Contest, The Nature of Time, December 1, 2008.

Barrett, Jonathan, and Nicolas Gisin. "How Much Measurement Independence Is Needed to Demonstrate Nonlocality?" *Physical Review Letters* 106, no. 10 (March 10, 2011).

Barz, Stefanie, Elham Kashefi, Anne Broadbent, Joseph F. Fitzsimons, Anton Zeilinger, and Philip Walther. "Demonstration of Blind Quantum Computing." *Science* 335, no. 6066 (January 20, 2012): 303–308.

Bauer, Henry H. *Scientific Literacy and the Myth of the Scientific Method*. Champaign, IL: University of Illinois Press, 1994.

Baum, Dan. "Deluged." *The New Yorker*, January 9, 2006.

Bekenstein, Jacob D. "Black Holes and Entropy." *Physical Review D* 7, no. 8 (April 15, 1973): 2333–46.

Bell, John S. "On the Einstein-Podolsky-Rosen Paradox." *Physics* 1, no. 3 (1964): 195–200.

———. *Speakable and Unspeakable in Quantum Mechanics*. 2nd ed. New York: Cambridge University Press, 2004.

Beller, Mara. *Quantum Dialogue: The Making of a Revolution*. Chicago: University of Chicago Press, 1999.

Beller, Mara, and Arthur Fine. "Bohr's Response to EPR." In *Niels Bohr and Contemporary Philosophy*, edited by Jan Faye and Henry J. Folse, 1–31. Boston Studies in the Philosophy of Science 153. Dordrecht: Springer Netherlands, 1994.

Belot, Gordon. "Understanding Electromagnetism." *The British Journal for the Philosophy of Science* 49, no. 4 (December 1998): 531–55.

Belousek, Darrin W. "Einstein's 1927 Unpublished Hidden-Variable Theory: Its Background, Context and Significance." *Studies in History and Philosophy of Science Part B* 27, no. 4 (December 1996): 437–61.

Berkson, William. *Fields of Force: The Development of a World View from Faraday to Einstein*. New York: Halsted Press/John Wiley and Sons, 1974.

Bern, Zvi, Lance J. Dixon, David C. Dunbar, and David A. Kosower. "Fusing Gauge Theory Tree Amplitudes into Loop Amplitudes." *Nuclear Physics B* 435, nos. 1–2 (February 6, 1995): 59–101.

Berryman, Sylvia. *The Mechanical Hypothesis in Ancient Greek Natural Philosophy*. New York: Cambridge University Press, 2009.

Black, Max. "Why Cannot an Effect Precede Its Cause?" *Analysis* 16, no. 3 (January 1956): 49–58.

Blum, Alexander. "From the Necessary to the Possible: The Genesis of the Spin-Statistics Theorem." *The European Physical Journal H* 39, no. 5 (December 2014): 543–74.

Bohm, David. "A Suggested Interpretation of the Quantum Theory in Terms of 'Hidden' Variables. I." *Physical Review* 85, no. 2 (January 1952): 166–79.

——. "A Suggested Interpretation of the Quantum Theory in Terms of 'Hidden' Variables. II." *Physical Review* 85, no. 2 (January 1952): 180–93.

Bohm, David, and Basil J. Hiley. *The Undivided Universe: An Ontological Interpretation of Quantum Theory.* London: Routledge, 1993.

Bohm, David, and J. P. Vigier. "Model of the Causal Interpretation of Quantum Theory in Terms of a Fluid with Irregular Fluctuations." *Physical Review* 96, no. 1 (October 1954): 208–16.

Bohr, Niels. "Discussion with Einstein on Epistemological Problems in Atomic Physics." In *Albert Einstein, Philosopher-Scientist*, edited by Paul Arthur Schilpp, 199–241. Chicago: Northwestern University Press, 1949.

——. *The Emergence of Quantum Mechanics (Mainly 1924–1926).* Edited by Klaus Stolzenburg. Vol. 5. New York: Elsevier, 1984.

——. *Foundations of Quantum Physics I (1926–1932).* Edited by Jørgen Kalckar. Vol. 6 of *Niels Bohr: Collected Works*, edited by Erik Rüdinger. New York: Elsevier, 1985.

——. *Foundations of Quantum Physics II (1933–1958).* Edited by Jørgen Kalckar. Vol. 7 of *Niels Bohr: Collected Works*, edited by Erik Rüdinger. New York: Elsevier, 1996.

——. "Space and Time in Nuclear Physics." In *Niels Bohr: Essays and Papers*, edited by John T. Sanders, translated by Else Mogensen. 1:205–20. Rochester, NY: Self-published, 1987.

Bombelli, Luca, Joohan Lee, David Meyer, and Rafael D. Sorkin. "Space-Time as a Causal Set." *Physical Review Letters* 59, no. 5 (August 3, 1987): 521–24.

Borisov, Alexey V., and Nikolay A. Kudryashov. "Paul Painlevé and His Contribution to Science." *Regular and Chaotic Dynamics* 19, no. 1 (February 9, 2014): 1–19.

Bork, Alfred M. "Maxwell and the Vector Potential." *Isis* 58, no. 2 (Summer 1967): 210–22.

Born, Max, and Albert Einstein. *The Born-Einstein Letters 1916–1955: Friendship, Politics and Physics in Uncertain Times.* Translated by Irene Born. New York: Palgrave Macmillan, 2005.

Bousso, Raphael. "The Holographic Principle." *Reviews of Modern Physics* 74, no. 3 (July–September 2002): 825–74.

Bouwmeester, Dik, Jian-Wei Pan, Klaus Mattle, Manfred Eibl, Harald Weinfurter, and Anton Zeilinger. "Experimental Quantum Teleportation." *Nature* 390 (December 11, 1997): 575–79.

Branciard, Cyril, Eric G. Cavalcanti, Stephen P. Walborn, Valerio Scarani, and Howard M. Wiseman. "One-Sided Device-Independent Quantum Key Distribution: Security, Feasibility, and the Connection with Steering." *Physical Review A* 85, no. 1 (January 3, 2012).

Brassard, Gilles. "Quantum Communication Complexity (A Survey)." *arXiv.org*, January 1, 2001.

Brassard, Gilles, Harry Buhrman, Noah Linden, André Allan Méthot, Alain Tapp, and Falk Unger. "Limit on Nonlocality in Any World in Which Communication Complexity Is Not Trivial." *Physical Review Letters* 96, no. 25 (June 27, 2006).

Britto, Ruth, Freddy Cachazo, Bo Feng, and Edward Witten. "Direct Proof of the Tree-Level Scattering Amplitude Recursion Relation in Yang-Mills Theory." *Physical Review Letters* 94, no. 18 (May 10, 2005).

Broderick, Avery E., Abraham Loeb, and Ramesh Narayan. "The Event Horizon of Sagittarius A*." *The Astrophysical Journal* 701, no. 2 (July 31, 2009): 1357–66.

Brush, Stephen G. "The Chimerical Cat: Philosophy of Quantum Mechanics in Historical Perspective." *Social Studies of Science* 10, no. 4 (November 1980): 393–447.

Cairncross, Frances. *The Death of Distance: How the Communications Revolution Will Change Our Lives*. Boston: Harvard Business School Press, 1997.

Caldiero, Frank M. "The Source of Hamlet's 'What a Piece of Work Is a Man!'" *Notes and Queries* 196 (September 29, 1951): 421–24.

Callan, Curtis G., Jr., Steven B. Giddings, Jeffrey A. Harvey, and Andrew Strominger. "Evanescent Black Holes." *Physical Review D* 45, no. 4 (February 15, 1992): 1005–1009.

Cantor, Geoffrey N. *Optics After Newton: Theories of Light in Britain and Ireland, 1704–1840*. Dover, NH: Manchester University Press, 1983.

Capra, Fritjof. *The Tao of Physics: An Exploration of the Parallels Between Modern Physics and Eastern Mysticism*. 3rd ed. Boston: Shambhala, 1991.

Carazza, Bruno, and Helge Kragh. "Heisenberg's Lattice World: The 1930 Theory Sketch." *American Journal of Physics* 63, no. 7 (July 1995): 595–605.

Carlip, Steven. "Spontaneous Dimensional Reduction?" In *The Sixth International School on Field Theory and Gravitation*, edited by Waldyr Alves Rodrigues Jr., Richard Kerner, Gentil O. Pires, and Carlos Pinheiro, 63–72. AIP Conference Proceedings 1483. Melville, NY: American Institute of Physics, 2012.

Carroll, Sean M. "In What Sense Is the Early Universe Fine-Tuned?" *arXiv.org*, June 11, 2014.

———. Opening talk at Setting Time Aright, conference of the Foundational Questions Institute, Bergen, Norway, and Copenhagen, Denmark, August 27–September 1, 2011.

———. *Spacetime and Geometry: An Introduction to General Relativity*. San Francisco: Addison-Wesley, 2004.

Casey, Edward S. *The Fate of Place: A Philosophical History*. Berkeley: University of California Press, 1997.

Castellani, Elena. "Dirac on Gauges and Constraints." *International Journal of Theoretical Physics* 43, no. 6 (June 2004): 1503–14.

Chambers, R. G. "Shift of an Electron Interference Pattern by Enclosed Magnetic Flux." *Physical Review Letters* 5, no. 1 (July 1, 1960): 3–5.

Chew, Geoffrey F. "The Dubious Role of the Space-Time Continuum in Subatomic Physics." *Science Progress* 51, no. 204 (October 1963): 529–39.

———. "'Bootstrap': A Scientific Idea?" *Science* 161, no. 3843 (August 23, 1968): 762–65.

Clifton, Robert K., and Hans P. Halvorson. "Entanglement and Open Systems in

Algebraic Quantum Field Theory." *Studies in History and Philosophy of Science Part B* 32, no. 1 (March 2001): 1–31.

Clifton, Robert K., D. V. Feldman, Hans P. Halvorson, Michael Redhead, and Alexander Wilce. "Superentangled States." *Physical Review A* 58, no. 1 (July 1, 1998): 135–45.

Coleman, Sidney. "Why There Is Nothing Rather Than Something: A Theory of the Cosmological Constant." *Nuclear Physics B* 310, nos. 3–4 (December 12, 1988): 643–68.

Collins, Harry M., and Trevor J. Pinch. *Frames of Meaning: The Social Construction of Extraordinary Science*. London: Routledge and Kegan Paul, 1982.

Copenhaver, Brian P. "Natural Magic, Hermetism, and Occultism in Early Modern Science." In *Reappraisals of the Scientific Revolution*, edited by David C. Lindberg and Robert S. Westman, 261–301. New York: Cambridge University Press, 1990.

Cornford, F. M. "The Invention of Space." In *The Concepts of Space and Time: Their Structure and Their Development*, edited by Milič Čapek, 3–16. Boston Studies in the Philosophy of Science 22. New York: Springer, 1976.

Cramer, John G. "The Quantum Handshake." The Alternate View, *Analog Science Fiction and Fact*, November 1986.

Craps, Ben, Savdeep Sethi, and Erik P. Verlinde. "A Matrix Big Bang." *Journal of High Energy Physics* 10, no. 1 (October 2005).

Cushing, James T. "Locality/Separability: Is This Necessarily a Useful Distinction?" *PSA: Proceedings of the Biennial Meeting of the Philosophy of Science Association* 1994. Vol. 1, *Contributed Papers*, 107–16. Chicago: University of Chicago Press, 1994.

———. *Quantum Mechanics: Historical Contingency and the Copenhagen Hegemony*. Chicago: University of Chicago Press, 1994.

———. *Theory Construction and Selection in Modern Physics: The S Matrix*. New York: Cambridge University Press, 1990.

Cushing, James T., and Ernan McMullin, eds. *Philosophical Consequences of Quantum Theory: Reflections on Bell's Theorem*. Notre Dame, IN: University of Notre Dame Press, 1989.

Davies, Paul C. W., and J. R. Brown. *The Ghost in the Atom: A Discussion of the Mysteries of Quantum Physics*. New York: Cambridge University Press, 1986. Reprint, 1993.

Davis, Tamara M., and Charles H. Lineweaver. "Expanding Confusion: Common Misconceptions of Cosmological Horizons and the Superluminal Expansion of the Universe." *Publications of the Astronomical Society of Australia* 21, no. 1 (2004): 97–109.

de Haas-Lorentz, Geertruida Luberta. *H. A. Lorentz: Impressions of His Life and Work*. Amsterdam: North Holland Publishing, 1957.

de Muynck, Willem M. "Can We Escape From Bell's Conclusion That Quantum Mechanics Describes a Non-Local Reality?" *Studies in History and Philosophy of Science Part B* 27, no. 3 (September 1996): 315–30.

Descartes, René. *Principles of Philosophy*. Translated by Valentine Rodger Miller and Reese P. Miller. New York: Springer, 1984.

———. *The Philosophical Writings of Descartes*. Vol. 3, *The Correspondence*. Translated by John Cottingham, Robert Stoothoff, and Dugald Murdoch. New York: Cambridge University Press, 1991.

d'Espagnat, Bernard. "The Quantum Theory and Reality." *Scientific American*, November 1979.

DeWitt, Bryce S. "Quantum Mechanics and Reality." *Physics Today* 23, no. 9 (September 1970): 155–65.

Dickens, Charles. *The Personal History of David Copperfield*. Edited by Trevor Blount, New York: Penguin Classics, 1966.

Dickson, Michael. *Quantum Chance and Non-Locality: Probability and Non-Locality in the Interpretations of Quantum Mechanics*. Cambridge: Cambridge University Press, 1998.

Dijksterhuis, E. J. *The Mechanization of the World Picture: Pythagoras to Newton*. Translated by C. Dikshoorn. Princeton, NJ: Princeton University Press, 1986.

Dirac, Paul Adrien Maurice. "Classical Theory of Radiating Electrons." *Proceedings of the Royal Society A: Mathematical and Physical Sciences* 167, no. 929 (August 5, 1938): 148–69.

———. "Quantised Singularities in the Electromagnetic Field." *Proceedings of the Royal Society A: Mathematical Physical and Engineering Sciences* 133, no. 821 (September 1, 1931): 60–72.

———. "The Evolution of the Physicist's Picture of Nature." *Scientific American*, May 1963, 45–53.

Dobbs, B.J.T. "Newton's Alchemy and His Theory of Matter." *Isis* 73, no. 4 (December 1982): 511–28.

Doran, Barbara Giusti. "Origins and Consolidation of Field Theory in Nineteenth-Century Britain: From the Mechanical to the Electromagnetic View of Nature." *Historical Studies in the Physical Sciences* 6 (1975): 133–260.

Dowker, Fay. "Causal Sets and the Deep Structure of Spacetime." In *One Hundred Years of Relativity: Space-Time Structure; Einstein and Beyond*, edited by Abhay Ashtekar, 445–64. Hackensack, NJ: World Scientific, 2005.

Dreyer, Olaf. "The World Is Discrete." *arXiv.org*, July 23, 2013.

Duff, Michael J. "The Theory Formerly Known as Strings." *Scientific American*, February 1998.

Duncan, Anthony. *The Conceptual Framework of Quantum Field Theory*. New York: Oxford University Press, 2012.

Earman, John. "Locality, Nonlocality, and Action at a Distance: A Skeptical Review of Some Philosophical Dogmas." In *Kelvin's Baltimore Lectures and Modern Theoretical Physics: Historical and Philosophical Perspectives*, edited by Robert Kargon and Peter Achinstein, 449–90. Cambridge, MA: The MIT Press, 1987.

———. *A Primer on Determinism*. Boston: Kluwer Academic Publishers, 1986.

———. "Recent Work on Time Travel." In *Time's Arrows Today: Recent Physical and Philosophical Work on the Direction of Time*, edited by Steven F. Savitt, 268–310. New York: Cambridge University Press, 1995.

———. *World Enough and Space-Time: Absolute Versus Rational Theories of Space and Time*. Cambridge, MA: The MIT Press, 1989.

Easther, Richard, Raphael Flauger, Paul McFadden, and Kostas Skenderis. "Constraining Holographic Inflation with WMAP." *Journal of Cosmology and Astroparticle Physics*, no. 9 (September 2011).

Ehlers, Jürgen. "The Nature and Structure of Spacetime." In *The Physicist's Conception of Nature*, edited by Jagdish Mehra, 71–91. Boston: D. Reidel Publishing Company, 1973.

Einstein, Albert. "Autobiographical Notes." In *Albert Einstein, Philosopher-Scientist*, edited by Paul Arthur Schilpp. New York: MJF Books, 1949.

———. *Einstein's Miraculous Year: Five Papers That Changed the Face of Physics*. Edited by John Stachel. Princeton, NJ: Princeton University Press, 1998.

———. Letter to Erwin Schrödinger, June 19, 1935, Old Lyme, CT. Einstein Archives Online, EA 22-47.

———. Letter to Paul Epstein, November 10, 1945, Princeton, NJ, EA 10-583.

———. *Letters to Solovine: 1906–1955*. New York: Philosophical Library, 1987. Reprint, New York: Open Road, 2011.

———. *The Meaning of Relativity*. Translated by Edwin Plimpton Adams. 5th ed. Princeton, NJ: Princeton University Press, 2005. First published 1922.

———. "On a Heuristic Point of View Concerning the Production and Transformation of Light." In Stachel, *Einstein's Miraculous Year*, 177–98.

———. "On the Electrodynamics of Moving Bodies." In Stachel, *Einstein's Miraculous Year*, 121–60.

———. "On the Quantum Theory of Radiation." In *Sources of Quantum Mechanics*, edited by B. L. van der Waerden, 63–77. Mineola, NY: Dover Publications, 2007. First published 1967.

———. "Physics and Reality." *Journal of the Franklin Institute* 221, no. 3 (March 1936): 349–82.

———. "Quanten-Mechanik Und Wirklichkeit." *Dialectica* 2, nos. 3–4 (November 1948): 320–24.

———. "Über die Entwicklung unserer Anschauungen über das Wesen und die Konstitution der Strahlung." *Physikalische Zeitschrift* 10, no. 22 (1909): 817–25.

———. "Über die vom Relativitätsprinzip geforderte Trägheit der Energie." *Annalen Der Physik* 328, no. 7 (1907): 371–84.

Einstein, Albert, and Nathan Rosen. "The Particle Problem in the General Theory of Relativity." *Physical Review* 48, no. 1 (July 1935): 73–77.

Einstein, Albert, Boris Podolsky, and Nathan Rosen. "Can Quantum-Mechanical Description of Physical Reality Be Considered Complete?" *Physical Review* 47, no. 10 (May 15, 1935): 777–80.

Ekert, Artur K. "Quantum Cryptography Based on Bell's Theorem." *Physical Review Letters* 67, no. 6 (August 5, 1991): 661–63.

Erler, Theodore G., and David J. Gross. "Locality, Causality, and an Initial Value Formulation for Open String Field Theory." *arXiv.org*, June 22, 2004.

Faraday, Michael. *Experimental Researches in Chemistry and Physics*. London, 1859.

———. *Experimental Researches in Electricity*. Vol. 1. London, 1839.

———. *Experimental Researches in Electricity*. Vol. 2. London, 1844.

———. *Experimental Researches in Electricity*. Vol. 3. London, 1855.

Fernbach, Philip M., Todd Rogers, Craig R. Fox, and Steven A. Sloman. "Political Extremism Is Supported by an Illusion of Understanding." *Psychological Science* 24, no. 6 (June 7, 2013): 939–46.

Feynman, Richard P. *QED: The Strange Theory of Light and Matter*. Princeton, NJ: Princeton University Press, 1986.

————. *The Character of Physical Law*. Cambridge, MA: The MIT Press, 1967.

————. *"Surely You're Joking, Mr. Feynman!": Adventures of a Curious Character*. New York: W. W. Norton, 1985. Reprint, 1997.

Fine, Arthur. "Do Correlations Need to Be Explained?" In Cushing and McMullin, *Philosophical Consequences of Quantum Theory*, 175–94.

————. *The Shaky Game: Einstein, Realism, and the Quantum Theory*. 2nd ed. Chicago: University of Chicago Press, 1996.

Finkelstein, David Ritz. "Space-Time Code." *Physical Review* 184, no. 5 (August 25, 1969): 1261–71.

Fletcher, Neville H. "Nonlinear Dynamics and Chaos in Musical Instruments." In *Complex Systems: From Biology to Computation*, edited by David G. Green and Terry Bossomaier, 106–17. Amsterdam: IOS Press, 1993.

Forman, Paul. "Weimar Culture, Causality, and Quantum Theory, 1918–1927: Adaptation by German Physicists and Mathematicians to a Hostile Intellectual Environment." *Historical Studies in the Physical Sciences* 3 (1971): 1–115.

Franson, J. D. "Generation of Entanglement Outside of the Light Cone." *Journal of Modern Optics* 55, no. 13 (July 20, 2008): 2117–40.

Fraser, Doreen. "The Fate of 'Particles' in Quantum Field Theories with Interactions." *Studies in History and Philosophy of Science Part B* 39, no. 4 (November 2008): 841–59.

Frazer, James George. *The Golden Bough: A Study in Magic and Religion. Part I: The Magic Art and the Evolution of Kings*. 3rd ed. Vol. 1. New York: Macmillan, 1913.

Freire, Olival, Jr. *The Quantum Dissidents: Rebuilding the Foundations of Quantum Mechanics (1950–1990)*. New York: Springer, 2015.

Friedman, John, Michael S. Morris, Igor Dmitrievich Novikov, Fernando Echeverria, Gunnar Klinkhammer, Kip S. Thorne, and Ulvi Yurtsever. "Cauchy Problem in Spacetimes with Closed Timelike Curves." *Physical Review D* 42, no. 6 (September 15, 1990): 1915–30.

Friedman, Michael. Introduction to *Metaphysical Foundations of Natural Science*, by Immanuel Kant, vii–xxx. Translated by Michael Friedman. New York: Cambridge University Press, 2004.

Frisch, Mathias. "Inconsistency in Classical Electrodynamics." *Philosophy of Science* 71, no. 4 (October 2004): 525–49.

————. "Non-Locality in Classical Electrodynamics." *The British Journal for the Philosophy of Science* 53 (2002): 1–19.

Gale, George. "Leibniz and Some Aspects of Field Dynamics." *Studia Leibnitiana* 6, no. 1 (1974): 28–48.

Galilei, Galileo. *Dialogue Concerning the Two Chief World Systems, Ptolemaic and Copernican*. Translated by Stillman Drake. Berkeley: University of California Press, 1953. Reprint, New York: Modern Library, 2001.

Galvez, Enrique. *Correlated-Photon Experiments for Undergraduate Labs*. Unpublished handbook, Colgate University, March 31, 2010.

Garber, Daniel. *Descartes' Metaphysical Physics*. Chicago: University of Chicago Press, 1992.

Gell-Mann, Murray. *The Quark and the Jaguar: Adventures in the Simple and the Complex*. New York: W. H. Freeman and Company, 1994. Reprint, New York: Henry Holt and Company, 1995.

Gell-Mann, Murray, and James B. Hartle. "Quantum Mechanics in the Light of Quantum Cosmology." In *Proceedings of the Third International Symposium Foundations of Quantum Mechanics in the Light of New Technology, Tokyo, August 28–31, 1989,* edited by Shun'ichi Kobayashi and Nihon Butsuri Gakkai, 321–43. Tokyo: Physical Society of Japan, 1990.

Gibbons, Gary W. "Black Holes and Information." Bits, Branes, Black Holes, Santa Barbara, CA, March 26, 2012.

Giddings, Steven B. "Black Holes, Quantum Information, and Unitary Evolution." *Physical Review D* 85, no. 12 (June 27, 2012).

Giddings, Steven B., Donald Marolf, and James B. Hartle. "Observables in Effective Gravity." *Physical Review D* 74, no. 6 (September 18, 2006).

Gilbert, William. *On the Magnet.* Translated by Silvanus P. Thompson. London: Chiswick Press, 1900.

Gilder, Louisa. *The Age of Entanglement: When Quantum Physics Was Reborn.* New York: Alfred A. Knopf, 2008.

Gisin, Nicolas. "Can Relativity Be Considered Complete? From Newtonian Nonlocality to Quantum Nonlocality and Beyond." *arXiv.org,* December 20, 2005.

Goodrick-Clarke, Nicholas. *The Western Esoteric Traditions: A Historical Introduction.* New York: Oxford University Press, 2008.

Gorelik, Gennadiĭ E. "Matvei Bronstein and Quantum Gravity: 70th Anniversary of the Unsolved Problem." *Physics-Uspekhi* 48, no. 10 (2005): 1039–53.

Goss, W. M., and R. X. McGee. "The Discovery of the Radio Source Sagittarius A (Sgr A)." In *The Galactic Center,* edited by Roland Gredel, 369–79. Astronomical Society of the Pacific Conference Series 102. San Francisco: Astronomical Society of the Pacific, 1996.

Gödel, Kurt. "A Remark About the Relationship Between Relativity Theory and Idealistic Philosophy." In Schilpp, *Albert Einstein, Philosopher-Scientist,* 555–62.

Greenberger, Daniel M., Michael A. Horne, Abner Shimony, and Anton Zeilinger. "Bell's Theorem Without Inequalities." *American Journal of Physics* 58, no. 12 (December 1990): 1131–43.

Greenblatt, Stephen. "The Answer Man." *The New Yorker,* August 11, 2011.

Greene, Brian. *The Fabric of the Cosmos: Space, Time, and the Texture of Reality.* New York: Vintage Books, 2005.

Gribbin, John. "The Man Who Proved Einstein Was Wrong." *New Scientist,* no. 1744, November 24, 1990.

Griffiths, Robert B. "EPR, Bell, and Quantum Locality." *American Journal of Physics* 79, no. 9 (September 2011): 954–65.

———. "Quantum Locality." *Foundations of Physics* 41, no. 4 (April 2011): 705–33.

Grim, Patrick. "Free Will in Context: A Contemporary Philosophical Perspective." *Behavioral Sciences and the Law* 25, no. 2 (March/April 2007): 183–201.

Gross, David J. "Einstein and the Search for Unification." *Current Science* 89, no. 12 (December 25, 2005): 2035–40.

———. "Twenty Five Years of Asymptotic Freedom." *Nuclear Physics B—Proceedings Supplements* 74, no. 1 (March 1999): 426–46.

Grünbaum, Adolf. *Modern Science and Zeno's Paradoxes.* Middletown, CT: Wesleyan University Press, 1967.

Guth, Alan H., and Paul J. Steinhardt. "The Inflationary Universe." *Scientific American*, May 1984.

Guthrie, W.K.C. *A History of Greek Philosophy*. Vol. 1, *The Earlier Presocratics and the Pythagoreans*. Cambridge: Cambridge University Press, 1962. Reprint, 1979.

———. *A History of Greek Philosophy*. Vol. 2, *The Presocratic Tradition From Parmenides to Democritus*. Cambridge: Cambridge University Press, 1965. Reprint, 1979.

Hall, Michael J. W. "Relaxed Bell Inequalities and Kochen-Specker Theorems." *Physical Review A* 84 (August 2, 2011).

Halvorson, Hans P. "Algebraic Quantum Field Theory." In *Philosophy of Physics*, edited by Jeremy Butterfield and John Earman, Part A, 731–864. Amsterdam: North Holland, 2007.

———. "Locality, Localization, and the Particle Concept: Topics in the Foundations of Quantum Field Theory," PhD diss., University of Pittsburgh, 2001.

Halvorson, Hans P., and Robert K. Clifton. "No Place for Particles in Relativistic Quantum Theories?" *Philosophy of Science* 69, no. 1 (March 2002): 1–28.

Hamma, Alioscia, and Fotini Markopoulou. "Background-Independent Condensed Matter Models for Quantum Gravity." *New Journal of Physics* 13, no. 9 (September 2011).

Hanegraaff, Wouter J. "The New Age Movement and the Esoteric Tradition." In *Gnosis and Hermeticism from Antiquity to Modern Times*, edited by Roelof van den Broek and Wouter J. Hanegraaff, 359–82. Albany: State University of New York Press, 1998.

Haroche, Serge, and Jean-Michel Raimond. *Exploring the Quantum: Atoms, Cavities, and Photons*. New York: Oxford University Press, 2006.

Harrison, Edward. *Cosmology: The Science of the Universe*. 2nd ed. New York: Cambridge University Press, 2000.

Hashimoto, Akikazu, and N. Itzhaki. "Traveling Faster Than the Speed of Light in Noncommutative Geometry." *Physical Review D* 63, no. 12 (May 21, 2001).

Hatfield, Brian. *Quantum Field Theory of Point Particles and Strings*. Boulder, CO: Westview Press, 1992.

Hawking, Stephen W. "Black Holes and Thermodynamics." *Physical Review D* 13, no. 2 (January 15, 1976): 191–97.

———. "Information Loss in Black Holes." *Physical Review D* 72, no. 8 (October 18, 2005).

He, Guang Ping. "Simple Quantum Protocols for the Millionaire Problem with a Semi-Honest Third Party." *International Journal of Quantum Information* 11, no. 2 (March 2013).

Healey, Richard A. *Gauging What's Real: The Conceptual Foundations of Contemporary Gauge Theories*. New York: Oxford University Press, 2007.

———. "Holism and Nonseparability." *The Journal of Philosophy* 88, no. 8 (August 1991): 393–421.

———. "New Thoughts on Yang-Mills Theories." Workshop on the Foundations of Gauge Theories, University of California, Irvine, March 21–23, 2014.

Heemskerk, Idse, Joao Penedones, Joseph Polchinski, and James Sully. "Holography from Conformal Field Theory." *Journal of High Energy Physics* 2009, no. 10 (October 2009).

Hegerfeldt, Gerard C. "Violation of Causality in Relativistic Quantum Theory?" *Physical Review Letters* 54, no. 22 (June 3, 1985): 2395–98.

Heller, Michael. "Where Physics Meets Metaphysics." In *On Space and Time*, edited by Shahn Majid, 238–77. New York: Cambridge University Press, 2008.

Heller, Michael, and Wiesław Sasin. "Einstein-Podolski-Rosen Experiment from Noncommutative Quantum Gravity." In *Particles, Fields, and Gravitation*, edited by Jakub Remblienski, 234–41. AIP Conference Proceedings 453. Woodbury, NY: American Institute of Physics, 1998.

———. "Emergence of Time." *Physics Letters A* 250, no. 1 (1998): 48–54.

———. "Nonlocal Phenomena From Noncommutative Pre-Planckian Regime." *arXiv.org*, June 17, 1999.

Henry, John. "Gravity and *De Gravitatione*: The Development of Newton's Ideas on Action at a Distance." *Studies in History and Philosophy of Science Part A* 42, no. 1 (2011): 11–27.

———. "Occult Qualities and the Experimental Philosophy: Active Principles in Pre-Newtonian Matter Theory." *History of Science* 24 (1986): 335–81.

———. "'Pray Do Not Ascribe That Notion to Me': God and Newton's Gravity." In *The Books of Nature and Scripture: Recent Essays on Natural Philosophy, Theology, and Biblical Criticism in the Netherlands of Spinoza's Time and the British Isles of Newton's Time*, edited by James E. Force and Richard Henry Popkin, 123–47. International Archives of the History of Ideas 139. Boston: Kluwer Academic Publishers, 1994.

———. *The Scientific Revolution and the Origins of Modern Science*. 2nd ed. New York: Palgrave Macmillan, 2002.

Henson, Joe. "The Causal Set Approach to Quantum Gravity." In Oriti, *Approaches to Quantum Gravity*, 393–413.

Hertz, Heinrich. *Miscellaneous Papers*. Edited by Philipp Lenard. Translated by Daniel Evan Jones and G. A. Schott. New York: Macmillan, 1896.

Hesse, Mary B. *Forces and Fields: The Concept of Action at a Distance in the History of Physics*. New York: Thomas Nelson, 1961. Reprint, Mineola, NY: Dover Publications, 2005.

Holbrow, C. H., Enrique Galvez, and M. E. Parks. "Photon Quantum Mechanics and Beam Splitters." *American Journal of Physics* 70, no. 3 (March 2002): 260–65.

Horowitz, Gary T., and Joseph Polchinski. "Gauge/Gravity Duality." In Oriti, *Approaches to Quantum Gravity*, 169–86.

Horowitz, Gary T., Albion Lawrence, and Eva Silverstein. "Insightful D-Branes." *Journal of High Energy Physics*, no. 7 (July 2009).

Horwich, Paul. *Asymmetries in Time: Problems in the Philosophy of Science*. Cambridge, MA: The MIT Press, 1987.

Hossenfelder, Sabine. "Disentangling the Black Hole Vacuum." *Physical Review D* 91, no. 4 (February 15, 2015).

Howard, Don. "Albert Einstein as a Philosopher of Science." *Physics Today* 58, no. 12 (December 2005): 34–40.

———. "Einstein on Locality and Separability." *Studies in History and Philosophy of Science Part A* 16, no. 3 (September 1985): 171–201.

———. "Holism, Separability, and the Metaphysical Implications of the Bell Experiments." In Cushing and McMullin, *Philosophical Consequences of Quantum Theory*, 224–53.

———. "'Nicht Sein Kann Was Nicht Sein Darf,' or the Prehistory of EPR, 1909–1935: Einstein's Early Worries About the Quantum Mechanics of Composite Systems." In *Sixty-Two Years of Uncertainty: Historical, Philosophical, and Physical Inquiries Into the Foundations of Quantum Mechanics*, edited by Arthur I. Miller, 61–111. New York: Plenum Press, 1990.

———. "A Peek Behind the Veil of Maya." In *The Cosmos of Science: Essays of Exploration*, edited by John Earman and John D. Norton, 87–150. Pittsburgh: University of Pittsburgh Press, 1997.

———. "Revisiting the Einstein-Bohr Dialogue." *Iyyun: The Jerusalem Philosophical Quarterly* 56 (January 2007): 57–90.

———. "*The Shaky Game: Einstein, Realism, and the Quantum Theory* by Arthur Fine." *Synthese* 86, no. 1 (January 1991): 123–41.

Huggett, Nick. "Zeno's Paradoxes." *Stanford Encyclopedia of Philosophy*, October 15, 2010.

Hume, David. *Of the Understanding*. Book 1 of *A Treatise of Human Nature*. Edited by L. A. Selby-Bigge. Oxford: Clarendon Press, 1888.

Hunt, Bruce J. *The Maxwellians*. Ithaca, NY: Cornell University Press, 1991. Reprint, 2005.

Hutchison, Keith. "What Happened to Occult Qualities in the Scientific Revolution?" *Isis* 73, no. 2 (June 1, 1982): 233–53.

Isaacson, Walter. *Einstein: His Life and Universe*. New York: Simon and Schuster, 2007.

Ismael, Jenann. "Decision and the Open Future." In *The Future of the Philosophy of Time*, edited by Adrian Bardon, 149–68. London: Routledge, 2011.

———. "What Entanglement Might Be Telling Us." Paper presented at the 46th Chapel Hill Colloquium in Philosophy, Chapel Hill, NC, November 2–4, 2012.

Jackson, John David. *Classical Electrodynamics*. 2nd ed. New York: John Wiley and Sons, 1975.

Jacobson, Ted. "Introduction to Quantum Fields in Curved Spacetime and the Hawking Effect." In *Lectures on Quantum Gravity*, edited by Andrés Gomberoff and Donald Marolf, 39–89. Series of the Centro De Estudios Científicos. New York: Springer, 2005.

Jaeger, Gregg, and Alexander V. Sergienko. "Multi-photon quantum interferometry." *Progress in Optics* 42 (2001): 277–324.

James, William. *The Will to Believe: And Other Essays in Popular Philosophy*. New York, 1897.

Jammer, Max. *Concepts of Force: A Study in the Foundations of Dynamics*. Cambridge, MA: Harvard University Press, 1957. Reprint, Mineola, NY: Dover Publications, 1999.

———. *Concepts of Space: The History of Theories of Space in Physics*. Foreword by Albert Einstein. 3rd, enlarged ed. Mineola, NY: Dover Publications, 1993. First published 1954.

———. *The Philosophy of Quantum Mechanics: The Interpretations of Quantum Mechanics in Historical Perspective*. New York: John Wiley and Sons, 1974.

Janiak, Andrew. *Newton as Philosopher*. New York: Cambridge University Press, 2008.

Jennewein, Thomas, Christoph Simon, Gregor Weihs, Harald Weinfurter, and Anton Zeilinger. "Quantum Cryptography with Entangled Photons." *Physical Review Letters* 84, no. 20 (May 15, 2000).

Jensen, Kristan, and Andreas Karch. "Holographic Dual of an Einstein-Podolsky-Rosen Pair Has a Wormhole." *Physical Review Letters* 111, no. 21 (November 21, 2013).

Johnson, George. "A Passion for Physical Realms, Minute and Massive." *The New York Times*, February 20, 2001.

Johnson, Steven. *Where Good Ideas Come From: The Natural History of Innovation*. New York: Riverhead Books, 2010.

Jones, Martin R., and Robert K. Clifton. "Against Experimental Metaphysics." *Midwest Studies in Philosophy* 18, no. 1 (September 1993): 295–316.

Kafatos, Menas, and Robert Nadeau. *The Conscious Universe: Part and Whole in Modern Physical Theory*. New York: Springer, 1990.

Kahneman, Daniel. *Thinking, Fast and Slow*. New York: Farrar, Straus and Giroux, 2011.

Kaiser, David. *Drawing Theories Apart: The Dispersion of Feynman Diagrams in Postwar Physics*. Chicago: University of Chicago Press, 2005.

———. *How the Hippies Saved Physics: Science, Counterculture, and the Quantum Revival*. New York: W. W. Norton, 2011.

Kant, Immanuel. *Critique of Pure Reason*. Edited by Vasilis Politis. London: J.M. Dent, 1993.

———. *Metaphysical Foundations of Natural Science*. Translated by Michael Friedman. New York: Cambridge University Press, 2004.

Kaplan, Marc. "Winners of the Young Researchers Competition in Physics Announced." Press release, Science and Ultimate Reality, meeting in honor of John Wheeler, Princeton, NJ, March 21, 2002.

Kaplunovsky, Vadim, and Marvin Weinstein. "Space-Time: Arena or Illusion?" *Physical Review D* 31, no. 8 (April 15, 1985): 1879–98.

Kavli Institute of Theoretical Physics. Bits, Branes, Black Holes, program, Santa Barbara, CA, March 19–May 25, 2012.

Kearney, Hugh. *Science and Change, 1500–1700*. New York: McGraw-Hill, 1971.

Kepler, Johannes. *New Astronomy*. Translated by William H. Donahue. New York: Cambridge University Press, 1992.

Kipnis, Naum. "Luigi Galvani and the Debate on Animal Electricity, 1791–1800." *Annals of Science* 44, no. 2 (March 1987): 107–42.

Klein, Martin J. "Einstein and the Wave-Particle Duality." *The Natural Philosopher* 3 (1964): 3–49.

Kochiras, Hylarie. "Gravity and Newton's Substance Counting Problem." *Studies in History and Philosophy of Science Part A* 40 (September 2009): 267–80.

Koestler, Arthur. *The Watershed: A Biography of Johannes Kepler*. New York: Anchor Books, 1960.

Konopka, Tomasz, Fotini Markopoulou, and Simone Severini. "Quantum Graphity: A Model of Emergent Locality." *Physical Review D* 77, no. 10 (May 27, 2008).

Kuchař, Karel V. "Time and Interpretations of Quantum Gravity." In *General Relativity and Relativistic Astrophysics: The 4th Canadian Conference, Winnipeg, Manitoba, Canada, 16–18 May 1991*, edited by Gabor Kunstatter, Dwight E. Vincent, and Jeff G. Williams. Hackensack, NJ: World Scientific, 1992.

Kuehn, Manfred. *Kant: A Biography*. New York: Cambridge University Press, 2001.

Kuhlmann, Meinard. "What Is Real?" *Scientific American*, August 2013, 40–47.

Kuhn, Thomas S. *The Copernican Revolution: Planetary Astronomy in the Development of Western Thought*. Cambridge, MA: Harvard University Press, 1957.

———. *The Structure of Scientific Revolutions*. Chicago: University of Chicago Press, 1962.

Kwiat, Paul G., Edo Waks, Andrew G. White, Ian Appelbaum, and Philippe H. Eberhard. "Ultrabright source of polarization-entangled photons." *Physical Review A* 60, no. 2 (August 1999): R773–76.

Lam, Vincent. "Structural Aspects of Space-Time Singularities." In *The Ontology of Spacetime II*, edited by Dennis Dieks, 111–31. Philosophy and Foundations of Physics 4. New York: Elsevier, 2008.

Landsman, N. P. "When Champions Meet: Rethinking the Bohr-Einstein Debate." *Studies in History and Philosophy of Science Part B* 37, no. 1 (March 2006): 212–42.

Lange, Marc. *An Introduction to the Philosophy of Physics: Locality, Fields, Energy, and Mass*. Malden, MA: Blackwell Publishing, 2002.

Langevin, Paul. "L'évolution de l'espace et du temps." *Scientia* 10 (1911): 31–54.

Lapkiewicz, Radek, Peizhe Li, Christoph Schaeff, Nathan K. Langford, Sven Ramelow, Marcin Wieśniak, and Anton Zeilinger. "Experimental Non-Classicality of an Indivisible Quantum System." *Nature* 474, no. 7352 (June 23, 2011): 490–93.

Laudisa, Federico. "Non-Local Realistic Theories and the Scope of the Bell Theorem." *Foundations of Physics* 38, no. 12 (December 2008): 1110–32.

Lefschetz, Solomon. *Introduction to Topology*. Princeton, NJ: Princeton University Press, 1949.

Leibniz, Gottfried Wilhelm. *New Essays on Human Understanding*. Abridged ed. Translated and edited by Peter Remnant and Jonathan Bennett. New York: Cambridge University Press, 1982.

———. *Philosophical Papers and Letters: A Selection*. Edited and translated by Leroy E. Loemker. Vol. 1. Chicago: University of Chicago Press, 1956.

———. *Philosophical Papers and Letters: A Selection*. Edited and translated by Leroy E. Loemker. 2nd ed. Vol. 2. Boston: Kluwer Academic Publishers, 1989.

Leucippus and Democritus. *The Atomists: Leucippus and Democritus; Fragments*. Translated by Christopher C. W. Taylor. Toronto: University of Toronto Press, 1999.

Liberati, Stefano, Sebastiano Sonego, and Matt Visser. "Faster-Than-C Signals, Special Relativity, and Causality." *Annals of Physics* 298, no. 1 (May 25, 2002): 167–85.

Lieb, Elliott H., and Derek W. Robinson. "The Finite Group Velocity of Quantum Spin Systems." *Communications in Mathematical Physics* 28, no. 3 (September 1972): 251–57.

Lightman, Alan P. "Magic on the Mind: Physicists' Use of Metaphor." *The American Scholar* 58, no. 1 (Winter 1989): 97–101.

Lightman, Alan P., and Roberta Brawer. *Origins: The Lives and Worlds of Modern Cosmologists*. Cambridge, MA: Harvard University Press, 1990.

Lincoln, Don. "Proving Special Relativity: Episode 2." Physics in a Nutshell, *Fermilab Today*, April 4, 2014.

List, Christian. "Free Will, Determinism, and the Possibility of Doing Otherwise." *Noûs* 48, no. 1 (March 2014): 156–78.

Lloyd, Geoffrey Ernest Richard. *Early Greek Science: Thales to Aristotle*. London: Chatto and Windus, 1970. Reprint, New York: W. W. Norton, 1974.

Lodge, Oliver. *Modern Views of Electricity*. London, 1889.

Loll, Renate, Jan Ambjørn, and Jerzy Jurkiewicz. "The Universe from Scratch." *Contemporary Physics* 47, no. 2 (March 2006): 103–17.

Lowe, David A., Joseph Polchinski, Leonard Susskind, Lárus Thorlacius, and John Uglum. "Black Hole Complementarity Versus Locality." *Physical Review D* 52, no. 12 (December 15, 1995): 6997–7012.

Lucretius. *The Nature of Things*. Translated by A. E. Stallings, New York: Penguin Classics, 2007.

Lyre, Holger. "Holism and Structuralism in U(1) Gauge Theory." *Studies in History and Philosophy of Science Part B* 35, no. 4 (December 2004): 643–70.

Macchetto, F. Duccio, and Mark Dickinson. "Galaxies in the Young Universe." *Scientific American* (May 1997): 92–99.

Mach, Ernst. *History and Root of the Principle of the Conservation of Energy*. Translated by Philip E. B. Jourdain. Chicago: Open Court Publishing, 1911.

———. *The Science of Mechanics: A Critical and Historical Account of Its Development*. Translated by Thomas J. McCormack. 4th ed. Chicago: Open Court Publishing, 1919. First published 1893.

Malament, David B. "In Defense of Dogma: Why There Cannot Be a Relativistic Quantum Mechanics of (Localizable) Particles." In *Perspectives on Quantum Reality: Non-Relativistic, Relativistic, and Field-Theoretic*, edited by Rob Clifton, 1–10. Boston: Kluwer Academic Publishers, 1996.

Maldacena, Juan, and Leonard Susskind. "Cool Horizons for Entangled Black Holes." *Fortschritte Der Physik* 61, no. 9 (September 2013): 781–811.

Mandelstam, Stanley. "Quantum Electrodynamics Without Potentials." *Annals of Physics* 19, no. 1 (July 1962): 1–24.

Mangano, Michelangelo L., and Stephen J. Parke. "Multi-Parton Amplitudes in Gauge Theories." *Physics Reports* 200, no. 6 (February 1991): 301–67.

Markopoulou, Fotini. "Space Does Not Exist, So Time Can." *arXiv.org*, September 10, 2009.

Markopoulou, Fotini, and Robert Lawrence Kuhn. "Why Is the Universe So Breathtaking?" Webcast, *Closer to Truth*, produced by the Kuhn Foundation, August 15, 2013.

Markopoulou, Fotini, and Lee Smolin. "Causal Evolution of Spin Networks." *Nuclear Physics B* 508, nos. 1–2 (December 22, 1997): 409–30.

———. "Disordered Locality in Loop Quantum Gravity States." *Classical and Quantum Gravity* 24, no. 15 (August 7, 2007): 3813–23.

———. "Quantum Theory from Quantum Gravity." *Physical Review D* 70, no. 12 (December 23, 2004): 124029.

Marolf, Donald. "Discussion: Holography and Unitarity in Black Hole Evaporation," Bits, Branes, Black Holes, Santa Barbara, CA, April 27, 2012.

———. "Holographic Thought Experiments." *Physical Review D* 79, no. 2 (January 29, 2009).

———. "Holography Without Strings?" *Classical and Quantum Gravity* 31, no. 1 (January 2014).

———. "Unitarity and Holography in Gravitational Physics." *Physical Review D* 79, no. 4 (February 9, 2009).

Martinec, Emil J. "Evolving Notions of Geometry in String Theory." *Foundations of Physics* 43, no. 1 (January 2013): 156–73.

Martinec, Emil J., Daniel Robbins, and Savdeep Sethi. "Toward the End of Time." *Journal of High Energy Physics*, no. 8 (August 1, 2006).

Mates, Benson. *The Philosophy of Leibniz: Metaphysics and Language.* New York: Oxford University Press, 1986.

Mathur, Samir D. "The Information Paradox: A Pedagogical Introduction." *Classical and Quantum Gravity* 26, no. 22 (November 21, 2009).

Matteucci, Giorgio, and Giulio Pozzi. "New Diffraction Experiment on the Electrostatic Aharonov-Bohm Effect." *Physical Review Letters* 54, no. 23 (June 10, 1985): 2469–72.

Mattle, Klaus, Harald Weinfurter, Paul G. Kwiat, and Anton Zeilinger. "Dense Coding in Experimental Quantum Communication." *Physical Review Letters* 76, no. 25 (June 17, 1996): 4656–59.

Maudlin, Tim. "Buckets of Water and Waves of Space: Why Spacetime Is Probably a Substance." *Philosophy of Science* 60, no. 2 (June 1993): 183–203.

———. *New Foundations for Physical Geometry: The Theory of Linear Structures.* New York: Oxford University Press, 2014.

———. "Part and Whole in Quantum Mechanics." In *Interpreting Bodies: Classical and Quantum Objects in Modern Physics*, edited by Elena Castellani, 46–60. Princeton, NJ: Princeton University Press, 1998.

———. *Quantum Non-Locality and Relativity: Metaphysical Intimations of Modern Physics.* 2nd ed. Malden, MA: Blackwell Publishers, 2002.

———. "Special Relativity and Quantum Entanglement: How Compatible Are They?" Intersectional Symposium: The Concept of Reality in Physics, Dresden, Germany, March 16, 2011.

Maxwell, James Clerk. *A Treatise on Electricity and Magnetism.* Vol. 1. Oxford, 1873.

———. *The Scientific Papers of James Clerk Maxwell.* Edited by W. D. Niven. Vol. 1. New York: Dover Publications, 1965.

———. *The Scientific Papers of James Clerk Maxwell.* Edited by W. D. Niven. Vol. 2. New York: Dover Publications, 1965.

McCormmach, Russell. "Einstein, Lorentz, and the Electron Theory." *Historical Studies in the Physical Sciences* 2 (1970): 41–87.

———. "H. A. Lorentz and the Electromagnetic View of Nature." *Isis* 61, no. 4 (Winter 1970): 459–97.

McFadden, Paul, and Kostas Skenderis. "Observational Signatures of Holographic Models of Inflation." In *The Twelfth Marcel Grossmann Meeting on Recent Developments in Theoretical and Experimental General Relativity, Astrophysics and Relativistic Field Theories*, edited by Thibault Damour, Robert T. Jantzen, and Remo Ruffini, 2315–23. Hackensack, NJ: World Scientific, 2012.

McGoldrick, Monica, Randy Gerson, and Sueli S. Petry. *Genograms: Assessment and Intervention.* New York: W. W. Norton, 2008.

McMullin, Ernan. "The Explanation of Distant Action: Historical Notes." In Cushing and McMullin, *Philosophical Consequences of Quantum Theory*, 272–302.

———. "The Origins of the Field Concept in Physics." *Physics in Perspective* 4, no. 1 (February 2002): 13–39.

Mehlberg, Henry. *Time, Causality, and the Quantum Theory*. Vol. 1, *Essay on the Causal Theory of Time*. Vol. 2, *Time in a Quantized Universe*. Edited by Robert Sonné Cohen. Studies in the Philosophy of Science 19. Boston: Kluwer Academic Publishers, 1980.

Melfos, Vasilios, Bruno Helly, and Panagiotis Voudouris. "The Ancient Greek Names 'Magnesia' and 'Magnetes' and Their Origin from the Magnetite Occurrences at the Mavrovouni Mountain of Thessaly, Central Greece. A Mineralogical–Geochemical Approach." *Archaeological and Anthropological Sciences* 3 (2011): 165–72.

Mermin, N. David. "Is the Moon There When Nobody Looks? Reality and the Quantum Theory." *Physics Today* 38, no. 4 (April 1985): 38–47.

Meschini, Diego, Markku Lehto, and Johanna Piilonen. "Geometry, Pregeometry and Beyond." *Studies in History and Philosophy of Science Part B* 36, no. 3 (September 2005): 435–64.

Mielczarek, Jakub. "Asymptotic Silence in Loop Quantum Cosmology." In *Multiverse and Fundamental Cosmology*, edited by Mariusz P. Dąbrowski, Adam Balcerzak, and Tomasz Denkiewicz, 81–84. AIP Conference Proceedings 1514. Melville, NY: American Institute of Physics, 2013.

Milburn, Gerard J. *The Feynman Processor: Quantum Entanglement and the Computing Revolution*. New York: Basic Books, 1998.

Miller, Arthur I. *Albert Einstein's Special Theory of Relativity: Emergence (1905) and Early Interpretation (1905–1911)*. New York: Springer, 1997. First published 1981 by Addison-Wesley.

Minkowski, Hermann, with notes by Arnold Sommerfeld. "Raum Und Zeit." In *Das Relativitätsprinzip: Eine Sammlung von Abhandlungen*, edited by H. A. Lorentz, Albert Einstein, and Hermann Minkowski, 54–71. Fortschritte Der Mathematischen Wissenschaften in Monographien. Leipzig: Vieweg+Teubner Verlag, 1923.

Mischel, Theodore. "Pragmatic Aspects of Explanation." *Philosophy of Science* 33, nos. 1–2 (January 1966): 40–80.

Misner, Charles W. "Feynman Quantization of General Relativity." *Reviews of Modern Physics* 29, no. 3 (July 1957): 497–509.

———. "Mixmaster Universe." *Physical Review Letters* 22, no. 20 (May 19, 1969): 1071–74.

Misner, Charles W., and John Archibald Wheeler. "Classical Physics as Geometry." *Annals of Physics* 2, no. 6 (December 1957): 525–603.

Misner, Charles W., Kip S. Thorne, and John Archibald Wheeler. *Gravitation*. San Francisco: W. H. Freeman, 1973.

Mitchell, Edward Page. "The Clock That Went Backward." In *The Time Traveler's Almanac*, edited by Ann VanderMeer and Jeff VanderMeer, 450–59. New York: Tor, 2014.

Mitroff, Ian I. "Norms and Counter-Norms in a Select Group of the Apollo Moon Scientists: A Case Study of the Ambivalence of Scientists." *American Sociological Review* 39, no. 4 (August 1973): 579–95.

Morong, William, Alexander Ling, and Daniel Oi. "Quantum Optics for Space Platforms." *Optics and Photonics News* 23, no. 10 (2012): 42–49.

Morus, Iwan Rhys. *When Physics Became King.* Chicago: University of Chicago Press, 2005.

Musser, George. *The Complete Idiot's Guide to String Theory.* New York: Alpha Books, 2008.

———. "Could Time End?" *Scientific American,* September 2010.

———. "Forces of the World, Unite!" *Scientific American,* September 2004.

———. "George and John's Excellent Adventures in Quantum Entanglement." *Observations, Scientific American* blog, January 30, 2012.

———. "How to Build the World's Simplest Particle Detector." *Critical Opalescence, Scientific American* blog, October 15, 2012.

———. "How to Build Your Own Quantum Entanglement Experiment, Part 2 (of 2)." *Critical Opalescence,* February 14, 2013.

———. "What Happens to Google Maps When Tectonic Plates Move?" *Critical Opalescence,* November 11, 2013.

Nastase, Horatiu. "The RHIC Fireball as a Dual Black Hole." *arXiv.org,* January 10, 2005.

Needham, Joseph, and Wang Ling. *Science and Civilisation in China.* Vol. 4, Part 1. Cambridge: Cambridge University Press, 1962.

Nerlich, Graham. *The Shape of Space.* New York: Cambridge University Press, 1976.

Newstead, Anne G. J. "Aristotle and Modern Mathematical Theories of the Continuum." In *Aristotle and Contemporary Science,* edited by Demetra Sfendoni-Mentzou, J. Hattiangadi, and David M. Johnson, 2:113–29. Frankfurt: Peter Lang, 2001.

Newton, Isaac. *Isaac Newton: Philosophical Writings.* Edited by Andrew Janiak. New York: Cambridge University Press, 2004.

———. *The Mathematical Principles of Natural Philosophy.* 2 vols. Translated by Andrew Motte. London, 1803.

Newton, T. D., and Eugene Paul Wigner. "Localized States for Elementary Systems." *Reviews of Modern Physics* 21, no. 3 (1949): 400–406.

Nikogosyan, D. N. "Beta Barium Borate (BBO)." *Applied Physics A: Solids and Surfaces* 52, no. 6 (1991): 359–68.

Nishioka, Tatsuma, Shinsei Ryu, and Tadashi Takayanagi. "Holographic Entanglement Entropy: An Overview." *Journal of Physics A: Mathematical and Theoretical* 42, no. 50 (December 2, 2009).

Norton, John D. "Einstein's Investigations of Galilean Covariant Electrodynamics Prior to 1905." *Archive for History of Exact Sciences* 59, no. 1 (November 2004): 45–105.

Nozick, Robert. *Invariances: The Structure of the Objective World.* Cambridge, MA: Belknap Press, 2001.

O'Connell, A. D., M. Hofheinz, M. Ansmann, Radoslaw C. Bialczak, M. Lenander, Erik Lucero, M. Neeley, D. Sank. H. Wang, M. Weides, J. Wenner, John M. Martinis, and A. N. Cleland. "Quantum Ground State and Single-Phonon Control of a Mechanical Resonator." *Nature* 464, no. 7289 (March 17, 2010): 697–703.

OK Go. "This Too Shall Pass." YouTube, March 1, 2010.

Oriti, Daniele, ed. *Approaches to Quantum Gravity: Toward a New Understanding of Space, Time and Matter.* New York: Cambridge University Press, 2009.

Pais, Abraham. *Einstein Lived Here*. New York: Oxford University Press, 1994.

Pauli, Wolfgang. "The Connection Between Spin and Statistics." *Physical Review* 58, no. 8 (October 15, 1940): 716–22.

Penrose, Roger. *The Road to Reality: A Complete Guide to the Laws of the Universe*. New York: Alfred A. Knopf, 2005.

Perkins, Ceri, and Sergei Malyukov. "Cables: The 'Blood Vessels' of ATLAS." *ATLAS Experiment*, "ATLAS News," January 2008.

Peskin, Michael E., and Daniel V. Schroeder. *An Introduction to Quantum Field Theory*. Boulder, CO: Westview Press, 1995.

Pettit, Philip. *The Common Mind: An Essay on Psychology, Society, and Politics*. New York: Oxford University Press, 1993.

Planck Collaboration. "Planck 2013 Results. XVI. Cosmological Parameters." *Astronomy and Astrophysics* 571 (October 29, 2014).

———. "Planck Intermediate Results. XXX. The Angular Power Spectrum of Polarized Dust Emission at Intermediate and High Galactic Latitudes." *Astronomy and Astrophysics* (forthcoming). Published electronically September 22, 2014.

Popescu, Sandu. "Bell's Inequalities and Density Matrices: Revealing 'Hidden' Nonlocality." *Physical Review Letters* 74, no. 14 (April 3, 1995): 2619–22.

Popper, Karl R. "Bell's Theorem: A Note on Locality." In Vol. 1 of *Microphysical Reality and Quantum Formalism*, edited by Alwyn van der Merwe, Franco Selleri, and Gino Tarozzi, 413–17. Proceedings of the Eponymous Conference, Urbino, Italy, September 25–October 3, 1985. Boston: Kluwer Academic Publishers, 1988.

———. *Quantum Theory and the Schism in Physics*. Vol. 3 of *The Postscript to the Logic of Scientific Discovery*, edited by W. W. Bartley III. London: Routledge, 1992.

———. "Three Views Concerning Human Knowledge." Chap. 3 in *Conjectures and Refutations: The Growth of Scientific Knowledge*, 97–119. London: Routledge, 1965.

Powers, Thomas. *Heisenberg's War: The Secret History of the German Bomb*. New York: Alfred A. Knopf, 1993.

Prescod-Weinstein, Chanda, and Lee Smolin. "Disordered Locality as an Explanation for the Dark Energy." *Physical Review D* 80, no. 6 (September 3, 2009).

Price, Huw. *Time's Arrow and Archimedes' Point: New Directions for the Physics of Time*. New York: Oxford University Press, 1996.

Prutchi, David, and Shanni R. Prutchi. *Exploring Quantum Physics Through Hands-On Projects*. Hoboken, NJ: John Wiley and Sons, 2012.

Ramsey, Norman F. *Spectroscopy with Coherent Radiation: Selected Papers of Norman F. Ramsey*. Hackensack, NJ: World Scientific, 1998.

Randall, Lisa, and Raman Sundrum. "A Large Mass Hierarchy from a Small Extra Dimension." *arXiv.org*, May 4, 1999.

Rayleigh, Lord. "The Dynamical Theory of Gases and of Radiation." *Nature* 72, no. 1855 (May 18, 1905): 54–55.

Redhead, Michael. *Incompleteness, Nonlocality, and Realism: A Prolegomenon to the Philosophy of Quantum Mechanics*. New York: Oxford University Press, 1987.

———. "More Ado About Nothing." *Foundations of Physics* 25, no. 1 (January 1995): 123–37.

Redner, S. "Citation Statistics from More Than a Century of Physical Review." *arXiv .org*, July 27, 2004.

Retzker, Alex, J. I. Cirac, and Benni Reznik. "Detecting Vacuum Entanglement in a Linear Ion Trap." *Physical Review Letters* 94, no. 5 (February 2005).

Reznik, Benni. "Distillation of Vacuum Entanglement to EPR Pairs." arXiv.org, August 1, 2000.

Rickles, Dean. "AdS/CFT Duality and the Emergence of Spacetime." *Studies in History and Philosophy of Science Part B* 44, no. 3 (2013): 312–20.

———. *Symmetry, Structure, and Spacetime*. New York: Elsevier, 2008.

Riemann, Bernhard Georg. "On the Hypotheses Which Lie at the Bases of Geometry." *Nature* 8, nos. 183–84 (May 8, 1873): 14–17, 36–37.

Ritz, Walther. "Recherches critiques sur l'électrodynamique générale." *Annales de chimie et de physique* 13 (1908): 145–275.

Rohrlich, Daniel, and Sandu Popescu. "Nonlocality as an Axiom for Quantum Theory." *arXiv.org*, August 9, 1995.

Rosenfeld, Léon. "Niels Bohr in the Thirties: Consolidation and Extension of the Conception of Complementarity." In *Niels Bohr: His Life and Work as Seen by His Friends and Colleagues*, edited by Stefan Rozental, 114–36. Amsterdam: North Holland, 1967.

Rossi, Paolo. *Francis Bacon: From Magic to Science*. Translated by Sacha Rabinovitch. Chicago: University of Chicago Press, 1978.

Rovelli, Carlo. "'Forget Time.'" Essay written for the Foundational Questions Institute Essay Contest, The Nature of Time, August 24, 2008.

———. "Aristotle's Physics: A Physicist's Look." *arXiv.org*, December 14, 2013.

———. *Quantum Gravity*. New York: Cambridge University Press, 2004.

Rowling, J. K. *Harry Potter and the Sorcerer's Stone*. New York: Scholastic, 1998.

Rozali, Moshe. "Comments on Background Independence and Gauge Redundancies." *arXiv.org*, September 23, 2008.

Ruijsenaars, S.N.M. "On Newton-Wigner Localization and Superluminal Propagation Speeds." *Annals of Physics* 137, no. 1 (November 1981): 33–43.

Saari, Donald G., and Zhihong Xia. "Off to Infinity in Finite Time." *Notices of the AMS* 42, no. 5 (May 1995): 538–46.

Sabín, C., J. J. García-Ripoll, E. Solano, and J. León. "Dynamics of Entanglement via Propagating Microwave Photons." *Physical Review B* 81, no. 18 (May 2010).

Sachdev, Subir. "Strange and Stringy." *Scientific American*, January 2013, 44–51.

Sack, Robert David. "Magic and Space." *Annals of the Association of American Geographers* 66, no. 2 (1976): 309–21.

Safranski, Rüdiger. *Schopenhauer and the Wild Years of Philosophy*. Translated by Ewald Osers. London: Weidenfeld and Nicolson, 1989. Reprint, Cambridge, MA: Harvard University Press, 1991.

Saint Thomas Aquinas. *The Summa Theologica of Saint Thomas Aquinas*. Vol. 1. Translated by Fathers of the English Dominican Province. Revised by Daniel J. Sullivan. Great Books of the Western World 19. Chicago: Encyclopaedia Britannica, 1952.

Salmon, Wesley C. *Causality and Explanation*. New York: Oxford University Press, 1998.

Sarovar, Mohan, Akihito Ishizaki, Graham R. Fleming, and K. Birgitta Whaley.

"Quantum Entanglement in Photosynthetic Light-Harvesting Complexes." *Nature Physics* 6 (April 25, 2010): 462–67.

Schachner, Nat. "Ancestral Voices." *Astounding Stories*, December 1933.

Schilpp, Paul Arthur, ed. *Albert Einstein: Philosopher-Scientist*. The Library of Living Philosophers, Vol. 7. Chicago: Northwestern University Press, 1949. Reprint, New York: MJF Books, 2001.

Schlick, Moritz. *General Theory of Knowledge*. Edited by Herbert Feigl. Translated by Albert E. Blumberg. Chicago: Open Court Publishing Company, 2002.

Schlosshauer, Maximilian, Johannes Kofler, and Anton Zeilinger. "A Snapshot of Foundational Attitudes Toward Quantum Mechanics." *Studies in History and Philosophy of Science Part B* 44, no. 3 (August 2013): 222–30.

Schmidt, Maarten. "3C 273: A Star-Like Object with Large Red-Shift." *Nature* 197, no. 4872 (March 16, 1963): 1040.

Schrödinger, Erwin. "Discussion of Probability Relations Between Separated Systems." *Mathematical Proceedings of the Cambridge Philosophical Society* 31, no. 4 (October 1935): 555–63.

———. "On Einstein's Gas Theory." *Physikalische Zeitschrift* 27 (1926): 95–101.

———. "Quantisation as a Problem of Proper Values (Part IV)." In *Collected Papers on Wave Mechanics*, 3rd (augmented) ed., 102–23. AMS Chelsea Publishing. Providence, RI: American Mathematical Society, 2003.

———. *Science and the Human Temperament*. Translated by James Murphy. London: George Allen and Unwin, 1935.

Schutz, Bernard. *Gravity from the Ground Up: An Introductory Guide to Gravity and General Relativity*. New York: Cambridge University Press, 2003.

Schwartz, H. M. "Einstein's Comprehensive 1907 Essay on Relativity, Part 1." *American Journal of Physics* 45, no. 6 (June 1977): 512–17.

Schweber, Silvan S. *QED and the Men Who Made It: Dyson, Feynman, Schwinger, and Tomonaga*. Princeton, NJ: Princeton University Press, 1994.

Seiberg, Nathan. "Emergent Spacetime." In *The Quantum Structure of Space and Time: Proceedings of the 23rd Solvay Conference on Physics, Brussels, Belgium, 1–3 December, 2005*, edited by David J. Gross, Marc Henneaux, and Alexander Sevrin, 163–78. Hackensack, NJ: World Scientific, 2007.

Sekino, Yasuhiro, and Leonard Susskind. "Fast Scramblers." *Journal of High Energy Physics* 2008, no. 10 (October 15, 2008).

Shimony, Abner. "Aspects of Nonlocality in Quantum Mechanics." In *Quantum Mechanics at the Crossroads: New Perspectives from History, Philosophy and Physics*, edited by James Evans and Alan S. Thorndike, 107–23. New York: Springer, 2007.

———. "Conceptual Foundations of Quantum Mechanics." In *The New Physics*, edited by Paul Davies, 373–95. New York: Cambridge University Press, 1989. Reprint, 1992.

Simplicius. *On Aristotle's Physics 6*. Ancient Commentators on Aristotle, edited by David Konstan. Ithaca, NY: Cornell University Press, 1989.

Sivers, Derek. "How to Start a Movement." TED Conference, Long Beach, CA, February 2010.

Slowik, Edward. "The Deep Metaphysics of Quantum Gravity: The Seventeenth Century Legacy and an Alternative Ontology Beyond Substantivalism and Relationism." *Studies in History and Philosophy of Science Part B* 44, no. 4 (November 2013): 490–99.

———. "The 'Properties' of Leibnizian Space: Whither Relationism?" *Intellectual History Review* 22, no. 1 (March 2012): 107–29.

Smerlak, Matteo, and Carlo Rovelli. "Relational EPR." *Foundations of Physics* 37, no. 3 (March 2007): 427–45.

Smith, D. Eric, and Duncan K. Foley. "Classical Thermodynamics and Economic General Equilibrium Theory." *Journal of Economic Dynamics and Control* 32, no. 1 (January 2008): 7–65.

Smolin, Lee. *The Life of the Cosmos.* New York: Oxford University Press, 1997.

———. "The Case for Background Independence." In *The Structural Foundations of Quantum Gravity,* edited by Dean Rickles, Steven French, and Juha T. Saatsi, 196–239. New York: Oxford University Press, 2006.

———. *Time Reborn: From the Crisis in Physics to the Future of the Universe.* New York: Houghton Mifflin Harcourt, 2013.

Sorkin, Rafael D. "Does Locality Fail at Intermediate Length-Scales?" In Oriti, *Approaches to Quantum Gravity,* 26–43.

Spekkens, Robert W. "The Paradigm of Kinematics and Dynamics Must Yield to Causal Structure." *arXiv.org,* August 31, 2012.

Stachel, John. *Einstein from 'B' to 'Z.'* Einstein Studies 9. The Center for Einstein Studies, Boston University. Boston: Birkhäuser, 2002.

Stamatellos, Giannis. *Plotinus and the Presocratics: A Philosophical Study of Presocratic Influences in Plotinus' Enneads.* Albany: State University of New York Press, 2007.

Stanley, H. E., L.A.N. Amaral, P. Gopikrishnan, P. Ch. Ivanov, T. H. Keitt, and V. Plerou. "Scale Invariance and Universality: Organizing Principles in Complex Systems." *Physica A: Statistical Mechanics and Its Applications* 281, nos. 1–4 (June 15, 2000): 60–68.

Stapp, Henry. "Space and Time in S-Matrix Theory." *Physical Review* 139, no. 1 (July 12, 1965): B257–70.

Stauffer, Robert C. "Speculation and Experiment in the Background of Oersted's Discovery of Electromagnetism." *Isis* 48, no. 1 (March 1957): 33–50.

Steinhardt, Paul J. "The Inflation Debate." *Scientific American,* April 2011.

Stone, A. Douglas. *Einstein and the Quantum: The Quest of the Valiant Swabian.* Princeton, NJ: Princeton University Press, 2013.

Strocchi, F. "Gauss' Law in Local Quantum Field Theory." In *Field Theory, Quantization and Statistical Physics: In Memory of Bernard Jouvet,* edited by E. Tirapegui, 227–36. Boston: Kluwer Academic Publishers, 1981.

———. "Relativistic Quantum Mechanics and Field Theory." *Foundations of Physics* 34, no. 3 (March 2004): 501–27.

Strominger, Andrew. "Inflation and the dS/CFT Correspondence." *Journal of High Energy Physics* 2001, no. 11 (November 2001).

Stuewer, Roger H. "The Experimental Challenge of Light Quanta." In *The Cambridge Companion to Einstein,* edited by Michel Janssen and Christoph Lehner, 143–66. New York: Cambridge University Press, 2014.

Stump, Eleonore, and Norman Kretzmann. "Eternity." *The Journal of Philosophy* 78, no. 8 (August 1981): 429–58.

Summers, Stephen J., and Reinhard Werner. "The Vacuum Violates Bell's Inequalities." *Physics Letters A* 110, no. 5 (July 29, 1985): 257–59.

Sundrum, Raman. "From Fixed Points to the Fifth Dimension." *Physical Review D* 86, no. 8 (October 2012).

Suppes, Patrick. "Descartes and the Problem of Action at a Distance." *Journal of the History of Ideas* 15, no. 1 (January 1954): 146–52.

Surowiecki, James. *The Wisdom of Crowds: Why the Many Are Smarter Than the Few and How Collective Wisdom Shapes Business, Economies, Societies, and Nations.* New York: Doubleday, 2004.

Susskind, Leonard. *The Cosmic Landscape: String Theory and the Illusion of Intelligent Design.* New York: Little, Brown, 2005.

Susskind, Leonard, and Edward Witten. "The Holographic Bound in Anti–de Sitter Space." *arXiv.org*, May 19, 1998.

Swingle, Brian. "Constructing Holographic Spacetimes Using Entanglement Renormalization." *arXiv.org*, September 14, 2012.

Tait, Peter Guthrie. *Properties of Matter.* Edinburgh, 1885.

Taylor, A. E. "Parmenides, Zeno, and Socrates." *Proceedings of the Aristotelian Society* 16 (1915): 234–89.

Taylor, Paul, Rich Morin, D'Vera Cohn, and Wendy Wang. "American Mobility: Who Moves? Who Stays Put? Where's Home?" Pew Research Center, December 29, 2008.

Tegmark, Max. "Parallel Universes." *Scientific American*, May 2003.

Teller, Paul. *An Interpretive Introduction to Quantum Field Theory.* Princeton, NJ: Princeton University Press, 1994.

't Hooft, Gerard. "Dimensional Reduction in Quantum Gravity." In *Salamfestschrift: A Collection of Talks From the Conference on Highlights of Particle and Condensed Matter Physics, ICTP, Trieste, Italy, 8–12 March 1993,* edited by Ahmed Ali, John Ellis, and S. Randjbar-Daemi, 284–96. Hackensack, NJ: World Scientific, 1994.

———. "Discreteness and Determinism in Superstrings." *arXiv.org*, July 16, 2012.

———. "The Future of Quantum Mechanics." Opening event, Emergent Quantum Mechanics 2013, 2nd International Symposium About Quantum Mechanics Based on a "Deeper Level Theory," Vienna, Austria, October 3–6, 2013.

Thucydides. *History of the Peloponnesian War.* Translated by Rex Warner. London: Penguin Books, 1954.

Thurschwell, Pamela. *Literature, Technology and Magical Thinking, 1880–1920.* New York: Cambridge University Press, 2001.

Tollaksen, Jeff, Yakir Aharonov, Aharon Casher, Tirzah Kaufherr, and Shmuel Nussinov. "Quantum Interference Experiments, Modular Variables and Weak Measurements." *New Journal of Physics* 12, no. 1 (January 2010).

Traweek, Sharon. *Beamtimes and Lifetimes: The World of High Energy Physicists.* Cambridge, MA: Harvard University Press, 1988. Reprint, 1992.

Trimmer, John D. "The Present Situation in Quantum Mechanics: A Translation of Schrödinger's 'Cat Paradox' Paper." *Proceedings of the American Philosophical Society* 124, no. 5 (October 10, 1980): 323–38.

Unruh, William G. "Minkowski Space-Time and Quantum Mechanics." In *Minkowski*

Spacetime: A Hundred Years Later, edited by Vesselin Petkov, 133–48. New York: Springer, 2010.

———. "Time, Gravity, and Quantum Mechanics." In Savitt, *Time's Arrows Today*, 23–65.

Ursin, Rupert, F. Tiefenbacher, T. Schmitt-Manderbach, H. Weier, Thomas Scheidl, M. Lindenthal, B. Blauensteiner, et al. "Entanglement-Based Quantum Communication Over 144 Km." *Nature Physics* 3 (July 2007): 481–86.

Vachaspati, Tanmay, and Mark Trodden. "Causality and Cosmic Inflation." *Physical Review D* 61, no. 2 (December 16, 1999).

Valentini, Antony. "Beyond the Quantum." *Physics World* 22, no. 11 (November 2009): 32–37.

van Fraassen, Bas C. *Quantum Mechanics: An Empiricist View*. New York: Oxford University Press, 1991.

van Lunteren, Frans H. "Framing Hypotheses: Conceptions of Gravity in the 18th and 19th Centuries." PhD diss., Utrecht University, 1991.

———. "Gravitational and Nineteenth-Century Physical Worldviews." In *Newton's Scientific and Philosophical Legacy*, edited by Paul B. Scheurer and G. Debrock. New York: Springer, 1988.

———. "Nicolas Fatio de Duillier on the Mechanical Cause of Universal Gravitation." In *Pushing Gravity: New Perspectives on Le Sage's Theory of Gravitation*, edited by Matthew R. Edwards, 41–59. Montréal: Apeiron, 2002.

Van Raamsdonk, Mark. "Building Up Spacetime with Quantum Entanglement." *General Relativity and Gravitation* 42, no. 10 (June 19, 2010): 2323–29.

Vedral, Vlatko. "High-Temperature Macroscopic Entanglement." *New Journal of Physics* 6 (August 9, 2004).

Verlinde, Erik P. "The Dark Phase Space of de Sitter," Bits, Branes, Black Holes, Santa Barbara, CA, April 3, 2012.

Von Plato, Jan. *Creating Modern Probability: Its Mathematics, Physics and Philosophy in Historical Perspective*. New York: Cambridge University Press, 1994.

Wald, Robert M. "Correlations and Causality in Quantum Field Theory." In *Quantum Concepts in Space and Time*, edited by Roger Penrose and C. J. Isham, 293–301. New York: Oxford University Press, 1986.

Wallace, David. *The Emergent Multiverse: Quantum Theory According to the Everett Interpretation*. New York: Oxford University Press, 2012.

Weatherall, James Owen. "The Scope and Generality of Bell's Theorem." *Foundations of Physics* 43, no. 9 (September 2013): 1153–69.

Weinberg, Steven. *Gravitation and Cosmology: Principles and Applications of the General Theory of Relativity*. New York: John Wiley and Sons, 1972.

———. "The Search for Unity: Notes for a History of Quantum Field Theory." *Daedalus* 106, no. 4 (Autumn 1977): 17–35.

Wen, Xiao-Gang. "Topological Order: From Long-Range Entangled Quantum Matter to a Unified Origin of Light and Electrons." *ISRN Condensed Matter Physics* 2013 (2013).

Westfall, Richard Samuel. "Newton and Alchemy." In *Occult and Scientific Mentalities in the Renaissance*, edited by Brian Vickers, 315–35. New York: Cambridge University Press, 1984.

———. "Newton and the Hermetic Tradition." In *Science, Medicine, and Society in the Renaissance: Essays to Honor Walter Pagel*, edited by Allen G. Debus, 2:183–98. New York: Science History Publications, 1972.

———. *The Construction of Modern Science: Mechanisms and Mechanics*. New York: Cambridge University Press, 1971.

———. *The Life of Isaac Newton*. New York: Cambridge University Press, 1993.

Wheeler, John Archibald, and Richard P. Feynman. "Classical Electrodynamics in Terms of Direct Interparticle Action." *Reviews of Modern Physics* 21, no. 3 (July–September 1949): 425–33.

Whitaker, Andrew. "John Bell in Belfast: Early Years and Education." In *Quantum (Un)Speakables: From Bell to Quantum Information*, edited by R. A. Bertlmann and A. Zeilinger, 7–20. New York: Springer, 2002.

Wijsman, Ellen M., and Luigi Luca Cavalli-Sforza. "Migration and Genetic Population Structure with Special Reference to Humans." *Annual Review of Ecology and Systematics* 15 (November 1984): 279–301.

Wilczek, Frank. "Quantum Field Theory." *Reviews of Modern Physics* 71, no. 2 (March 1999): S85–S95.

Williams, Leslie Pearce. *Michael Faraday: A Biography*. New York: Da Capo Press, 1971.

———. *The Origins of Field Theory*. Lanham, MD: University Press of America, 1980.

Wilson, Robert Woodrow. "The Cosmic Microwave Background Radiation." Nobel Lecture, Stockholm, December 8, 1978.

Wiseman, Howard M., S. J. Jones, and A. C. Doherty. "Steering, Entanglement, Nonlocality, and the Einstein-Podolsky-Rosen Paradox." *Physical Review Letters* 98, no. 14 (April 6, 2007).

Witten, Edward. "Anti–de Sitter Space, Thermal Phase Transition, and Confinement in Gauge Theories." *Advances in Theoretical and Mathematical Physics* 2, no. 3 (1998): 505–32.

———. "An Interpretation of Classical Yang-Mills Theory." *Physics Letters B* 77, nos. 4–5 (August 28, 1978): 394–98.

———. "Perturbative Gauge Theory as a String Theory in Twistor Space." *Communications in Mathematical Physics* 252, nos. 1–3 (December 2004): 189–258.

———. "Reflections on the Fate of Spacetime." *Physics Today* 49, no. 4 (April 1996): 24–30.

Wright, Jessey. "Quantum Field Theory: Motivating the Axiom of Microcausality." Master's thesis, University of Waterloo, 2012.

Wu, Tai Tsun, and Chen Ning Yang. "Concept of Nonintegrable Phase Factors and Global Formulation of Gauge Fields." *Physical Review D* 12, no. 12 (December 15, 1975): 3845–57.

Yates, Frances A. *Giordano Bruno and the Hermetic Tradition*. Chicago: University of Chicago Press, 1964. Reprint, 1991.

Zahar, Elie. "Why did Einstein's Programme Supersede Lorentz's? (II)." *British Journal for the Philosophy of Science* 24, no. 3 (September 1973): 223–62.

Zee, Anthony. *Quantum Field Theory in a Nutshell*. Princeton, NJ: Princeton University Press, 2010.

Zeilinger, Anton. "On the Interpretation and Philosophical Foundation of Quantum Mechanics." In *Vastakohtien Todellisuus: Festschrift for Professor K. V. Lauri-*

kainen, edited by Urho Ketvel et al., 167–78. Helsinki: Helsinki University Press, 1996.

———. "Testing Concepts of Reality with Entangled Photons in the Laboratory and Outside." Intersectional Symposium: The Concept of Reality in Physics, Dresden, Germany, March 16, 2011.

Zel'dovich, Yakov B. "Particle Production in Cosmology." *Journal of Experimental and Theoretical Physics Letters* 12, no. 9 (November 10, 1970): 307–11.

Zimbardo, Philip. *The Lucifer Effect: Understanding How Good People Turn Evil.* New York: Random House, 2008.

Acknowledgments

One of the reasons I love writing about physics is that physicists and philosophers of physics are so generous with their time and so tolerant of endless questions. I have many of them to thank for reading sections of my manuscript: Nima Arkani-Hamed, John Baez, Julian Barbour, Raphael Bousso, Sean Carroll, Artur Ekert, Enrique Galvez, Alan Guth, Hans Halvorson, Alioscia Hamma, John Henry, Sabine Hossenfelder, Don Howard, Nick Huggett, Jenann Ismael, David Kaiser, Meinard Kuhlmann, Fotini Markopoulou, Donald Marolf, Emil Martinec, Lionel Mason, Tim Maudlin, Chad Orzel, Joe Polchinski, Huw Price, Dean Rickles, Carlo Rovelli, Moshe Rozali, Kostas Skenderis, Brian Swingle, Jeff Tollaksen, David Tong, Jaroslav Trnka, Mark Van Raamsdonk, and David Wallace.

I had the great fortune of spending fall 2011 at the Centre for Quantum Technologies at the National University of Singapore, courtesy of the director, Artur Ekert, and the outreach manager, Jenny Hogan. The Foundational Questions Institute provided a mini-grant to help pay my expenses there, for which Theiss Research handled the mini-logistics. The Kavli Institute for Theoretical Physics hosted me in spring 2012, a visit made possible by then director David Gross. I'm eternally grateful to my bosses at *Scientific American*, Mariette Di Christina, Fred Guterl, and Ricki Rusting, for not throwing a brick at me when I asked to go on these mini-sabbaticals. To the contrary, they were enthusiastic on my behalf, even though my time out of the office couldn't have made their jobs any easier.

I'm also thankful to several other organizations for travel support: the Nordic Institute for Theoretical Physics and the Foundational Questions Institute, for travel to the "Nonlocality: Aspects and Consequences" workshop in Stockholm in June 2012; the Fetzer Franklin Fund, for travel to the "Emergent Quantum Mechanics" symposium in Vienna in October 2013; and the Department of Logic and Philosophy of Science at UC Irvine, for travel to the "Foundations of Gauge Theories" conference there in March 2014.

I'd never have powered through the trying middle stages of the project were it not for pep talks from David Biello, Lee Billings, Steve Colyer, Amanda Gefter, David Grinspoon, Leslie Mullen, and Luba Ostashevsky. Not only were they supportive, their own writings have never ceased to inspire me.

Jen Christiansen, a *Scientific American* colleague and friend, prepared many of the diagrams, and did so with characteristic flair and efficiency. Adrianne Mathiowetz shot the jacket photo after much exploration of the MIT campus for the best site.

Amanda Moon is the book editor that every writer hopes for: attentive and supportive, enthusiastic about the project, and full of ideas to make it better. Her assistants, Laird Gallagher and Scott Borchert, the production team, Elizabeth Gordon and Nina Frieman, and Debra Helfand kept the logistics humming as smoothly as a clockwork universe. And whoa, what a cover—I have the cover designer Jennifer Carrow to thank for that. Annie Gottlieb must have some kind of x-ray vision, to judge from her penetrating copyediting. Brian Gittis has been great at setting up a book tour, and thanks also to the book designer, Jonathan Lippincott. I'm also fortunate that the big bang was set up in such a way as to propel me into my agent Susan Rabiner's orbit.

Special thanks to Le Petit Parisien and Trend Coffee and Tea House, the two cafes in Montclair, New Jersey, where I spent many a day metabolizing caffeine and almond-chocolate croissants into physics.

By being a good sport when I had to stay in and write rather than go rock climbing or ice-skating, my daughter, Eliana, made the project possible. And when I was feeling down, she knew exactly what zombie-apocalypse allusion would restore my confidence. And no one has given me as visceral a feeling of space and time as my wife, Talia. I feel every mile between us and every moment apart. Oh, and she's the world's greatest human thesaurus. I dedicate this book to her and Eliana.

Index

Page numbers in *italics* refer to illustrations.